The
Great
WAVES

The Great WAVES

Douglas Myles

Foreword by George Pararas-Carayannis

McGRAW-HILL BOOK COMPANY

New York   St. Louis   San Francisco
Hamburg   Mexico   Toronto

Endpaper etching, "Earthquake at Lisbon in 1755,"
Mary Evans Picture Library, London, England.

1 2 3 4 5 6 7 8 9 D O C D O C 8 7 6 5

ISBN 0-07-044237-1

LIBRARY OF CONGRESS CATALOGING IN PUBLICATION DATA

Myles, Douglas.
The great waves.
Bibliography; p.
Includes index.
1. Tsunamis.   2. Seismology.   I. Title.
GC221.2.M95   1985   551.47'022   85-5913
ISBN 0-07-044237-1

Book design by Kathryn Parise

For my mother,
ISABEL NIBECKER,
who kept the faith,
and for
ELIZABETH MYLES,
who loved me in Eden

Acknowledgments

The author wishes to express his appreciation and indebtedness to Gordon D. Burton, Geophysicist in Charge, Pacific Tsunami Warning Center, Ewa Beach, Hawaii, for making available certain technical data that proved helpful in the preparation of Chapter 12.

Most in particular, my gratitude is given to George Pararas-Carayannis, Director, International Tsunami Information Center, Honolulu, Hawaii, who very kindly reviewed the manuscript, gave invaluable advice, and expressed his faith in the book by lending his name and prestige in the writing of the Foreword.

Finally, my profound thanks are given to Michael Hennelly, who discovered this book and took action to bring it to light, and to Thomas Quinn, whose subsequent faith in it made everything possible.

Contents

Foreword   *xi*

Introduction   *1*

1   *Earth Structure*   *5*

     Suppositions concerning earth's internal structure based on
seismology / How seismographs work

2   *The Topography of the Seas*   *13*

     The Pacific Ocean Basin / The Mendocino Fracture Zone and
others / Trenches and ridges / The theory of continental
drift / Theories of convection currents, contraction, and
expansion / A crack in the Earth 40,000 miles long

3   *Undersea Volcanic Eruptions and Submarine
Earthquakes*   *27*

     Krakatoa, "The Blast Heard Round the World!"—and the
waves it created / Lieutenant Billings's account of the end of
Arica / Tsunami on the open sea

4   *Alaska and the Hawaiian Islands*   *47*

     The Aleutian Trench: April 1, 1946 / Unimak Island and the
Scotch Cap Lighthouse / Hawaii: Hilo and the Waipio Valley /
Kawela Bay: Another eyewitness description / Observations
based on fluid dynamics

5   *The New Global Tectonics*   *65*

     Crustal plate slippage and disaster in South America / Chile:
May 21, 1960

Contents

6 *Alaska Again*   77
   Movement along the Denali Fault system: March 27, 1964

7 *Japan, the Ever-Vulnerable*   97
   The Japan Trench and the Tuscarora Deep / The state of
   Japanese seismicity / Scourge of the Sanriku Coast

8 *Tsunami of the Seventeenth and Eighteenth
   Centuries*   111
   Port Royal, Jamaica; Lisbon, Portugal

9 *Other Forms of Wave and Some Clear-Cut
   Distinctions*   129
      Seiches, bores, and whirlpools / "Seismic waves" as
      distinguished from "seismic sea waves" / The "ultimate
      wave"—Lituya Bay, July 9, 1958 / The phenomenon of
      swashes

10 *Ancient Disasters*   149
      Seismic instability of the eastern Mediterranean / The
      evidence of the spade: Heinrich Schliemann and Sir Arthur
      Evans / Santorini, Crete, and the Bronze Age destruction of
      the Minoan Empire / The greatest tsunami in the recorded
      history of man

11 *The Santorini Tsunami*   169
      The Hebrew Exodus from Egypt

12 *Can Man Defend Himself? The Future*   185
      The Pacific and Alaskan tsunami warning centers, their
      methods, instrumentation, and reach

    Bibliography   195

    Index   199

Foreword

Tsunami are among the most terrifying and complex physical phenomena. They have been responsible for great loss of lives and extensive destruction to property throughout mankind's history. Although infrequent, tsunami, because of their destructiveness, have had important impact on the human and socioeconomic sectors of our societies. The historical records show that enormous destruction of coastal communities throughout the world has been caused by such great waves.

The impact of tsunami on human societies can be traced back in written history to 1480 B.C. in the Eastern Mediterranean, when the Minoan civilization was wiped out by such great tsunami waves generated by the volcanic explosion of the island of Santorin. In the Pacific Ocean where the majority of these waves have been generated, the historical record, although brief, shows tremendous destruction. In Japan, which has one of the most populated coastal regions in the world and a long history of earthquake activity, tsunami have destroyed entire coastal communities. There is also history of tsunami destruction in Alaska, in the Hawaiian Islands, and in South America. North and South American records have dated such events back to 1788 for Alaska and 1562 for Chile. Records of Hawaiian tsunami go back to 1821.

While most of the destructive tsunami have occurred in the Pacific Ocean, devastating tsunami have also occurred in the Atlantic and Indian Oceans, as well as in the Mediterranean Sea. A large tsunami accompanied the earthquake of Lisbon in 1755, that of the Mona Passage of Puerto Rico in 1918, and at the Grand Banks of Canada in 1929.

In the last 20 to 30 years there has been tremendous growth and development of the coastal areas in most of the developing or developed Pacific nations. This is the result of a population explosion and of technological and economic developments that have made the use of the coastal zone more necessary than before. Many of the countries of the Pacific, for example, have populations with natural maritime orientation. For many of these countries, foreign trade is a necessity, so they maintain major port facilities. Others have extensive shipbuilding facilities, electric plants, refineries, and other important coastal structures. Similarly many of the other countries of the Pacific have harbors as bases for their fishing industries. Peru, for example, at the port of Callao, near Lima, maintains a large fleet for anchovy fishing. Callao is located near a strong seismic and potentially tsunamigenic region that from time to time has given rise to these great destructive waves.

Finally, a number of coastal sites throughout the Pacific have begun important aquacultural industries and canneries. This combination of social and economic factors makes a number of developed and developing Pacific Islands and continental Pacific nations socially and economically vulnerable to the threat of tsunami. The extensive coastal boundaries, the number of islands, the long coastlines of Pacific countries containing a number of vulnerable engineering structures, the numerous large ports, the productive fishing and aquaculture industries, and the great density of population in the coastal areas, make tsunami an ever-threatening natural hazard.

To protect life and property in the Pacific an International Pacific Tsunami Warning System has been organized and is presently functioning, utilizing an extensive network of seismic and tidal stations, as well as communications, to ensure that the warning information is prompt and accurate. This International Tsunami Warning Network is operating under the auspices of the Intergovernmental Oceanographic Commission (UNESCO) and works very closely with each participating Pacific country to assess the tsunami hazard and mitigate its impact. Also a program of preparedness has been developed to alert coastal populations and industries, and civil defense agencies have been instituted in communities to respond to tsunami warnings. An International Tsunami Information Center has been created which coordinates public educational programs for each participating country working closely with government agencies, private institutions, and civil defense authorities, for the purpose of developing sound coastal management policies, which include zoning and planning for tsunami-prone coastal areas.

Although numerous educational pamphlets and important scientific articles have been written on tsunami, the literature is relatively poor in providing descriptive information and accounts of historical events. This information was badly needed.

The present book, *The Great Waves*, will fulfill this need. It has been thoroughly and comprehensively researched, and compiles diverse materials on tsunami. The author has organized these materials into a very readable format and has produced a fine book which should appeal to scientists and laymen alike. The book provides the necessary background on the earth's structure, the underwater topography of the oceans, and the earth's violent events, then proceeds to the explanation of the tsunami itself, which is a geophysical as well as an oceanographic phenomenon.

Mr. Myles has done a fine job in documenting, in very explicit narrative form, important destructive historic tsunamis throughout the world from the beginning of recorded history to the present time. Many of these descriptive materials are based on eyewitness accounts of the different disastrous tsunami and have been painstakingly collected, organized, and presented in this book, providing not only descriptions of the tsunami disasters themselves, but also giving us an inside glimpse of the history of each particular era, and the ever-interesting personal accounts of people who were affected by these events. Thus *The Geat Waves* is a very readable and entertaining as well as an educational book, about one of the least understood natural disasters that has plagued our planet since its beginning.

—Dr. George Pararas-Carayannis
Director, International Tsunami
Information Center
Honolulu, Hawaii

The
Great
WAVES

Introduction

A little after midnight on Monday, August 23, 1976, an earthquake measuring eight points on the Richter scale, having its epicenter in the Celebes Sea, struck the southern Philippines. The temblor lasted only 20 seconds. However, in that brief length of time it generated what are usually called "tidal waves," causing them to move out in concentric circles at high speed and fall with crushing force on the defenseless southwest coast of Mindanao Island.

As such waves go, these were not very large. Reports coming out of the stricken area spoke of a consensus of 30 feet of height. But, striking while most of the inhabitants were still dazedly digging themselves out, they completed the earthquake devastation in the coastal communities of Alicia, Pagadian City, Cotabato, and Davao, left 90,000 people homeless, injured nearly 2,000, and killed almost 8,000, the majority by drowning.

The waves were far from unique. Their like had been seen before in many parts of the world over many centuries of time. Waves like them, indeed far larger and more destructive than they, will be seen again. And over and over again, ignorance of what they are and what they can do will play its part in the destruction of human habitation and human life.

What is a tidal wave? Admittedly, there remains today in lay circles a great deal of confusion about it, the product of neglect or misinformation. Characteristically, even the popular name, "tidal wave," is incorrectly applied, since the waves thus referred to have nothing to do with the tides. To the oceanographers, seismologists, and vulcanologists who have occasion to study them, these terrifying natural phenomena are called seismic sea waves or *tsunami*,

1

the latter being a Japanese word (the same in both the singular and plural) meaning "large waves in harbors." Again, an unfortunate usage, perhaps, for as a descriptive term it is not especially accurate. But it is acceptable to the world of science, as a tacit recognition of the frequency with which Japan has fallen victim.

There are many points of interest. How big do tsunami become? What is their nature and origin? Historically, what has been their effect? Need we fear them in the future? This book, will attempt to answer all these questions and many more. At this point, withholding detail for the body of the text, we can touch very briefly on the question of size.

Most people today have seen photographs of the waves that in winter roll in at world-famous Waimea Bay on the north shore of Oahu Island in the Hawaiian group. Some have actually stood and watched them. A tiny handful of expert and daring surfers from several countries have ridden them on boards. These combers are not tsunami, of course, but unusually large sea waves of the familiar type, acquiring their spectacular size from the effect of waves traversing deep water which suddenly encounter shallow reefs, and their speed from the absence of a continental shelf to slow them down. They are easily 45 feet in height, so large that a surfer attempting to "hang ten" just beneath the curl of one of them is reduced to gnat-size against the ponderous watery backdrop. Yet imagine a wave *a hundred feet high. Two hundred. Three hundred.* Imagine such a wave as once drowned whole islands and their populations, submerged continental coasts to a distance of more than a hundred miles inland, annihilated forever an entire civilization. In the Pacific and Mediterranean Seas especially, such waves *have* been generated in the geologic past, some within the brief span of man's recorded history. In any sea on earth they may be generated again. These are the greatest waves the world has ever known. These are tsunami.

Specifically, seismic sea waves may result from the following causes: (1) an upthrusting or downthrusting of the sea bed in an area where the earth's crust is unstable, resulting in a high magnitude earthquake; (2) a major undersea landslide; (3) a submarine volcanic eruption of a certain type and intensity. To understand how this can happen, and why, it is necessary to know something about the structure of the planet, particularly with regard to what are called *zones of shear* within the crust, and the behavior of water when subjected to enormous, very sudden changes of pressure. It is the author's intention to pursue this understanding a step at a time.

Earth Structure

Suppositions concerning earth's internal structure based on
seismology.   How seismographs work.

Having a circumference at the Equator of 24,902 miles, and a
diameter of 7,917 miles, the earth is very small as planets go, an
obloid sphere slightly flattened at its poles, weighing in the area of
6 billion trillion tons. It isn't all of the same composition throughout.
On the contrary, between the four main parts of the sphere there
are distinct and important differences. We have no provable facts
concerning the interior. Our assumptions, though products of the
best scientific thinking, are based entirely on conjecture and are
not amenable to conclusive testing. However, in one of its most
violent manifestations, nature has provided man with certain clues.
Virtually everything he has learned about the nonobservable inner
regions has come from his study of earthquakes.

For purposes of science, the names of the divisions of the Earth
are as follows: crust, mantle, outer core, and inner core. Let us
examine them one at a time.

Of the four delineated zones, the crust is by far the thinnest,
comprising 1½ percent of the earth's total mass. Under the six
continents, it may be more than 40 miles thick in places (with an
average thickness of 20 miles), and is basically light granite. Under
both seas and continents, it runs to a depth of from three to nine
miles, and is largely composed of heavier basalt. It therefore consists
of two separate layers, a lighter resting on a heavier. The lighter
or upper layer (called the *sial* because silicon and aluminum are
its principal elements) is superimposed on the lower and heavier

one (called *sima* because it is composed primarily of silicon and magnesium), The sial underlies much of the continental land masses. The foundation stone of the seas is mostly sima.

Beneath the crust, and extending all the way to the core, lies a zone of greater density called the mantle, some 1,800 miles thick, embracing 80 percent of the earth's volume. Unlike the crust, it does not appear to be rigid. In fact, the consensus of current scientific theory holds that it possesses the malleable characteristics of a plastic and *flows* as an extremely thick fluid. Very little is known about it. To date, geophysicists can only speculate concerning its composition. Periodotite, eclogite, and dunite, three rocks found on the surface, are all believed to exist in the mantle, with dunite thought by some to be in great abundance. One surmise for which some solid evidence exists is that circulatory movement within the viscous mantle plays an extremely important role in crustal changes (see theory of continental drift in Chapter 2), often disastrously for man. If this is so, then the heat responsible for that continuous motion would logically come from the core.

As has been noted, the core appears to consist of two parts, an inner and outer layer. Taken together, they are about 4,300 miles in diameter, which means that the radius from the exact center of the Earth to the point where the outer core joins the mantle is computed at 2,150 miles. This breaks down to a thickness of 1,300 miles for the outer core and 850 miles for the inner core. The majority opinion today is that the outer core consists mainly of molten iron, with possibly nickel and traces of cobalt, while the inner core, chemically quite similar, is both solid and hotter. One might well ask why the hotter of the two would not also be a fluid. The theoretical answer is based on pressure. Standing at sea level, a human is under a pressure of 1 atmosphere, or approximately 15 pounds to the square inch. The rock that exists at the point where the mantle and core join is under 16,950,000 pounds per square inch. The pressure at the core's lower levels is much greater, so great that the molecular displacement necessary for rock to flow cannot occur.

As the primary mineral from which the core is composed, why should iron be selected? In point of fact, its preponderant existence there is almost a certainty. Considering that its volume is only about 260 billion cubic miles, the earth is the heaviest planet in this solar system, with an average density 5.5 times greater than water. Something must account for this tremendous weight. The crust and mantle together could not begin to contribute markedly to it. Therefore, the core must. Its estimated 42 billion cubic miles of volume would logically be iron. This metal is common in the universe, the only extremely heavy one known to be. Under the estimated temperatures and pressures prevailing at the earth's core (not less than 588 pounds per cubic foot), the density of iron would satisfy the calculated weight.

It has already been noted that what little we know of the earth's interior we derive from seismological investigation, seismology being the study of earthquakes. For our purposes, we needn't dwell at length on the origins of this remarkable science. The earliest instrument for detecting earth tremors dates from the second-century Han Dynasty in China, when, in A.D. 132, the court astronomer, Chang Heng, invented a crude form of seismograph. Since that far-off day, there have been many improvements. But it was not until after the mid-nineteenth century that Palmieri, an Italian, and Milne, an Englishman, developed instruments of practical utility. Since his was the first device to actually measure ground movements, John Milne is credited with having founded the modern science of seismology.

Today, seismograph stations are located all over the world, many of them in each civilized country. How do they function? Basically, there are not less than three seismographs in any single laboratory, each mounted on very firm foundations, bolted in what are generally called piers. Often, the lab itself is an underground chamber. But whether it is or not, the metal frames enclosing the instruments are bonded in the bedrock below, their rods going deep enough to respond to all subterranean vibrations.

The seismographs work on the pendulum principle. But since a pendulum is insensitive to all motion transmissions except those in its own plane of oscillation, they must be set at right angles to each other, one seismograph to record vertical earth movements, and two others to record movements on the east-west and north-south axes.

In principle, a simple seismograph would consist of a weighted pendulum hung from its rock-anchored frame by a coil of wire. The pendulum is equipped with a recording device (such as a stylus writing on smoked paper, an ink-pen on ordinary paper, a fine jet of ink, a tracing burned electrically, or a beam of light focused on photographic paper), which traces the movements of an opposing frame and its attached instrument as the earth shakes. This tracing is made on lined graph paper moving constantly about a slowly revolving drum. (For example, on one type of seismograph the drum, marked off in minutes, revolves once each half hour, and the 11 lines on the paper represent a time-period of 5½ hours.) The line traced is usually nearly straight, but whenever ground movement occurs the characteristic zigzags appear. Depending on the type of seismograph used, they can record acceleration, speed, or displacement.

The sensitivity of some seismographs is extraordinary. With these very sophisticated models the line traced is invariably serrated, as the instrument records the minutest vibrations of a planet whose crust is never static. One might expect that storms at sea would leave their marks, or the pounding of a great waterfall, or an underground river, or man-made explosions, and they all do. But so does surf breaking on a shoreline, or even variations of barometric pressure on the earth's incredibly thin "skin." Today's electronic amplification is so acute that tremors can be magnified 100,000 times. Indeed, the claim has been made that the crawling of an insect inside a seismological vault can leave its tracing on a graph.

The tiniest irregularities, those whose tracings compose the vast majority of any seismological record, are called microseisms. Through a careful study of them, scientists have been known to detect the birth of a hurricane hundreds of miles distant. But they have their disadvantages, too. Unless they are sufficiently "damped" by instrumental adjustment, they have a tendency to encroach on the tracings left by earthquakes, thus confusing the readings.

We have said that man's suppositions regarding the earth's internal structure are derived largely from his study of earthquakes. When a quake strikes, shock waves are transmitted in all directions from its surface point of origin, known as the epicenter. Passing completely through the earth, they return to the surface in curving lines and can be recorded on seismographs thousands of miles away. They are basically of three types, each of which is slower than its predecessor. The first to arrive, called P (or primary) waves, are of comparatively low amplitude (height), and have a short period (distance from crest to crest). Like sound waves, they compress matter as they pass through it, and because their thrust is in the line of their own propagation they are also known as *longitudinal* waves. They penetrate all matter whatsoever, solid, liquid, or gas.

According to the distance from the quake's focus (depth of origination below the epicenter), seconds or minutes can pass before the second train of waves, the S (or secondary) waves, arrive at the seismograph station. Their amplitude is greater than that of P's and their period longer. Because their passage through the earth induces lateral rather than vertical vibration in rock masses, they are often called *shear* or *transverse* waves. They will pass only through solid substances. If gaseous or liquid matter were to intervene, they would be deflected or absorbed.

Arriving last, we have the L (or long) waves, their amplitude and period much greater still, their path sharply distinguished from those of the P's and S's in that it follows the longer line of the crust rather than passing through the mantle or core. The L waves proper (sometimes called "love's waves") should be distinguished from the M's (of maximum amplitude) and the C's (from the Latin *coda,* meaning tail), which come at the very end.

When completed, the tracings made on the recording paper are called a seismogram, and, as has been indicated, there is more than one kind. Basically, there are two methods of recording shock waves. The first is to record them on the same seismogram, first the P's, then the S's, then the L's, just as they arrive. The second method is to use two seismograms. Here, the graph paper of the first type has a great many finely spaced lines and is used to record P and S waves; i.e., waves with a period ranging between $1/10$ second and 5 seconds. The graph paper of the second type has far fewer lines more widely spaced and its instrument responds to the surface L waves; i.e., waves of a period of 8 to 600 seconds. Called *short-period* and *long-period* records, such seismograms are in general use today.

We have said that man's knowledge of the earth's interior, though still largely speculative, derives almost entirely from his study of earthquakes. This is based mainly on the travel times of shock waves from their epicenters to the recording devices around the world. The nearer the earth's center that P and S waves travel, the greater their speed. However, at certain known depths abrupt alterations of speed occur, indicating that the waves have encountered physical or chemical changes in the matter through which they pass or which has deflected them. These zones radically affecting the speed of such waves are called *discontinuities*. Two principal ones have been detected.

The discontinuity closest to the earth's surface bears the name of the Croatian seismologist who discovered it in 1909, Professor Andrija Mohorovičić. Studying the wave patterns of an earthquake in the Balkans, he found that P and S waves increased their speed at a depth beneath the surface of about twenty miles and continued to do so until they struck the core. Obviously, he reasoned, beginning at about that depth, the material through which they passed must be considerably more dense than that of the crust and hence must indicate a boundary, a dividing line between crust and mantle. This is the *Mohorovičić discontinuity*—called, for short, "the Moho."

Striking the Moho, P and S waves suddenly accelerate from 4.3 and 2.4 miles per second, respectively, to 5 and 2.9 miles per second. Proceeding downward through denser material, their speed steadily increases until, at a depth of 1,800 miles, it attains 8.5 and 4.5 miles per second, respectively. Here there is another abrupt change. Suddenly, as though again encountering an altogether different substance, the speed of P waves is reduced to 5 miles per second, and S waves simply vanish. The boundary between mantle and core has been reached. Named after another great seismologist, this second, far deeper line of division is called the *Gutenberg discontinuity*.

In both the crust and the mantle, lesser discontinuities are known to exist. Though as yet it bears no one's name, quite clearly, the one dividing the inner from the outer core is both well-defined and important.

Obviously, a knowledge of earth structure and the working principles of individual seismographs would be relatively pointless without an understanding of their capabilities. Taken singly, no seismograph can pinpoint the epicenter of an earthquake, but any one of them can accurately determine its distance from the recording station. This is done by noting the time interval between the arrival of P and S waves, and from this (with the speed of the waves provided by the Jeffreys-Bullen Table), computing the distance traveled. To illustrate, let's say we have three recording laboratories, since that is the minimum number. Using dividers, the scientists at each location project circles on a map, with the radii in each case representing the distance from the station to the source of the disturbance. The point of intersection of the three circles is the quake's epicenter.

Though this may sound very elementary, it should be pointed out that it is not always quite as neat and simple as this. When the focus of an earthquake is shallow, say, one to six miles deep in the crust, the arcs will normally intersect at the same point. However, occasionally the focus is deep, separated from the epicenter (the spot directly above on the surface) by 100 to 200 miles or even more. When this happens, the intersection of the arcs can be dispersed, sometimes widely so. Then, in order to make the arcs intersect at a given point, the depth of the disturbance must be calculated.

The procedure for this is well known. Depending on the distance of the recording station from the epicenter, one of three techniques is used. If the station is separated from the quake's point of origin by a large portion of the planet, then the deep wave passing from the focus *through the core* will appear on the seismogram considerably earlier than the surface wave from the epicenter. The difference in the elapsed transmission time of the two waves will give the depth of focus.

Secondly, in the case where the laboratory is fairly close to the epicenter, the procedure is to calculate the time difference between the wave *reflected from the core* and the surface wave.

Finally, for those stations located at positions intermediate to the two extremes, the measurement is between the wave *passing through the mantle* (but never touching the core) and the surface wave.

2

The Topography of the Seas

The Pacific Ocean Basin. The Mendocino Fracture Zone and others. Trenches and ridges. The theory of continental drift. Theories of convection currents, contraction, and expansion. A crack in the Earth 40,000 miles long.

At this writing, it seems evident that man will ultimately learn far more about the undersea world than has yet been discovered, that to date he has scarcely begun his quest for knowledge of this vast area. The long delay is understandable. Until early in this century the only means of taking soundings was by casting a lead line, exactly as ancient navies had done from their war galleys more that 2,000 years ago. While the centuries passed very little progress was made. Then, in 1773, the British Navy's Captain Constantine Phipps made the first successful sounding beyond the edge of the continental shelf, in 683 fathoms. And in 1840 Sir James Clark Ross became the first to plumb abyssal depths, at 2,425 fathoms.

These efforts, though milestones in the history of man's attempts to become better acquainted with his planet, did comparatively little to advance his knowledge of what lay beneath the water that covers seventy percent of it. A few more years passed. When, in 1866, "The Great Iron Ship," the *Great Eastern*, vindicated her much-ridiculed English designer by successfully laying the first trans-atlantic telegraph cable, public interest in the ocean depths quickened. Between 1872 and 1876, the circumnavigation of the globe by H.M.S. *Challenger* brought momentous changes. Using much-improved equipment, she took far more soundings than had ever been made before, supplied chart makers and biologists with invaluable information, and (not the least fascinating of her many signal discoveries) even gathered evidence for the existence of a strange and unknown rise in the middle of the Atlantic, the previously unguessed-at Mid-Atlantic Ridge. It was this memorable three-year voyage which laid the foundation of modern oceanography.

So far as the study of the sea floor is concerned, in the years immediately following World War I a major breakthrough was achieved. This was the echo sounder, a device for determining depth by timing the transmission of ultrasonic signals, or "pings," which reflect off the ocean bottom as a returning echo. This instrument has steadily improved. By the time World War II was under way profiles of the sea bottom could be continuously drawn by surface ships to an accuracy of 1 fathom in 3,000. And what an enormous difference it made. Now, within the short space of a few years, more was learned about the topography of the oceans than in all of man's previous history, and that topography was vastly unlike what had often been imagined. Generally, except for the abyssal plains, the contours of the ocean floor were found to be sharp, rugged, and well-defined, offering infinite and startling variety. Discovered there were gorges of an immensity to dwarf the greatest canyons on land, gigantic deltas of long-drowned rivers, soaring cliffs three thousand miles long—and all of it locked in the endless darkness of uncounted ages. Scientists were literally athirst for this heretofore untapped knowledge.

Within the depths of the sea, basically there are two rather well-defined levels. The first, the *continental slope,* beginning where the continents end, has an average depth of about 100 fathoms (600 feet) along its upper margin; the second, the *deep-ocean basins,* an average depth of between 2,500 and 3,000 fathoms (15,000 to 18,000 feet). Added to these is a comparatively shallow area marking the transition from land to sea, the *continental shelf.*

It is the last-named with which we have the greatest familiarity, on whose bounty we are dependent for most of the fish and shellfish we consume, all of the seaweed products, even many of the reserves of petroleum. Forming the coastline of most of the earth's land masses, the continental shelf extends seaward for a world average of 42 miles, and in the Atlantic in particular is often more than 100 miles wide. Although at places along the shores of the Barents Sea it attains a width of 750 miles, its greatest average of 400 miles is found along the coast of the Siberian Arctic. Most of the sea life observed and classified by man is found on the shelf, which falls away in gentle inclines to an average depth of approximately 600 feet. Between 300 and 600 feet, it is as a rule quite flat, a contour apparently the result of a combination of current and deep-wave action on the one hand and river-deposited sediments on the other. In several important aspects the shelf seems more closely identified with the land than with the sea, and indeed in many places offers evidence of once having been above water. Using explosives, deeply penetrating seismic surveying equipment has plumbed its depths in many areas. Beneath layers of ancient sediment these instruments have traced the outlines of beaches, bays, and river beds dating from earlier geologic periods millions of years ago.

At the outer edge of the shelf there is a sudden increase in depth of from 2 to 3 miles as the land falls away to form the *continental slope,* a decline that continues to the deep floor of the ocean. Though the slope extends out to sea for 40 to 50 miles in most areas, its gradient differs markedly from one region to another, just as do those of mountain ranges on land. As the boundary of the continents, even in its average global aspect such a slope is impressive. On some coasts, in some regions, it is more than that. Mountains sometimes rise very close to the ocean, leaving no space for a shelf between. That is the case where Aconcagua (called the "Butcher of the Andes" and the highest peak in that range) plunges to the Peru-Chile Trench. The base of the mountain is nearly five miles below the water's surface. The horizontal distance from the summit to the bottom of the trench is nearly 100 miles, the vertical drop over 9 miles. This is the largest, most spectular declivity on earth.

Thus far, the description of the slopes may have given the impression of a comparatively unrelieved decline presenting little in the way of topographical variation. This is far from so. As undersea probes have shown, the slopes are deeply fissured with countless submarine canyons, the origin of which is still a matter of controversy. Geologically speaking, they are quite young. Some have been traced to the Cenozoic Era, the most recent of geologic ages; most were likely formed in the Pleistocene Epoch, not over a million years ago. Earthquakes undoubtedly created some of them. But since they are too far out and too deep to be accounted for as submerged river beds, the present consensus points to turbidity currents. These undersea avalanches with the mild-sounding name are anything but mild. Silt piles up in the millions of tons, accumulating over the years until static pressure combines with gravitational pull to bring it down. When this occurs, the mud-laden flow comes rushing down the slope at great speed, cutting gorges through the layers of sediment. It may surge ahead for hundreds of miles before friction reduces its kinetic energy, causing it to fan out and eventually stopping it. Turbidity currents are suspected agents in the peacetime loss of submarines by several nations. At various times they have broken transoceanic telegraph cables, on at least one occasion (Grand Banks, Newfoundland, November 18, 1929), snapping a whole series of them in a time sequence commensurate with their advance.

The end of the *continental slope* is marked by the beginning of the *deep-ocean basin,* an unbelievably vast area encompassing half of the planet's total surface and five-sevenths of its sea-covered regions. Unlike the shelf, which at various times has been successively above water and submerged, geologists believe the basin has never known anything but submersion, and has therefore never been subject to the wearing-down effects of wind and ice erosion. This does not mean it hasn't seen change, for volcanic activity is universally present here; the thrusting and folding, the mountain-building, the great fissuring undersea earthquakes. But nothing is worn smooth as on the land. The towering cliffs, the awesome canyons, the titanic mountains of the basin are all sharply contoured, their silhouettes as harsh and jagged (after hundreds of millions of years) as on the day of their creation. Only in those few areas called abyssal plains is the ocean bottom flat. Here, where turbidity currents have laid down sediments over countless millennia, endlessly filling the depressions and planing them smooth, the order of levelness far exceeds that of the terrestrial plains. They are, in fact, the flattest expanses to be found on earth.

*   *   *

Imagine an area, roughly round in shape, of such colossal size that all the land masses, all six continents, could be fitted into it with considerable room remaining. That is the Pacific Ocean Basin, the largest, most varied and complex topographical entity on the planet. If its water could be removed it would be seen that its bottom (except for the abyssal plains) is extraordinarily irregular, for it contains the world's greatest chains of volcanoes, crustal fractures, and suboceanic trenches. Hawaii, the highest of all mountains (if measured from its base rather than from sea level), is located here, its 32,024 feet considerably higher than Everest's snow-crowned summit. Here, too, at the sourthern extremity of the Mariana Trench, is found the greatest depth in the crust, the 36,198-foot Challenger Deep.

Because it is rimmed by most of the world's volcanoes, many of which are still active, the Pacific Basin is often called by geologists the "Ring of Fire." Although this line of volcanic (and seismic) instability is periodically submerged as it passes beneath broad stretches of sea, in general it follows the continental shorelines in its eastern quadrants, and various island arcs to the west. To illustrate, we might begin tracing it at the Antarctic's Ross Sea, where the active volcano Erebus pinpoints it for us. From there it crosses the southwestern Pacific, passes through New Zealand, swings west just north of Australia, then north again through the Marianas, Japan, the Kuril Islands, and the Soviet Union's Kamchatka Peninsula, to join with the Aleutian chain. From the Aleutians it pursues a southerly course along the west coasts of North America, Mexico, Central and South America, ultimately again crossing the South Pacific to close on Antarctica at the Palmer Peninsula.

It should be added that in the western Pacific, above Australia, the Ring of Fire is laterally extended westward by a line of volcanoes and earth faults passing through Indonesia and reaching into the Andaman Sea almost to the Burmese coast. Geologically speaking, the very high degree of vulcanicity (eruptive volcanic activity) in this region makes it worthy of note. However, much of the vast area circumscribed above shares this explosive quality. A glance at Dr. Karl Sapper's Revised Vulcanicity Tables, the standard reference, shows Indonesia, Central America, and the Alaska-Aleutians complex as standing first, second, and third, respectively, in the worldwide "index of explosiveness," and all border the Pacific. In fact, of the more than 2,500 volcanic eruptions recorded in history, the Ring of Fire has been the scene of about 2,200.

But we mentioned more than volcanoes. We spoke of "crustal fractures" in the basin, and "sub-oceanic trenches." Even today, approximately 80 percent of the earthquakes that split or bulge the earth's crust occur within the Ring of Fire. This is not difficult to understand. To study any recently made seismic survey chart of the Pacific Ocean floor is to discover an area whose crust is incredibly unstable. Throughout two-thirds of its circumference, stretching all the way from the Aleutian Islands to the Indian Ocean, the eastern one-third of the Pacific Basin is ringed by fissures so numerous and extensive it gives the appearance of old rawhide which has lain long in the sun. These cracks are latitudinal, though on a Mercator projection they give the appearance of becoming gradually longitudinal as the basin curves and its waters merge with the Indian Ocean. How large are these crevasses? Seismic measurements show them to be anywhere from ten to sixty miles wide. Beginning close to the continental shoreline, they extend seaward on an average between 1,600 and 3,300 miles and have to date reached a maximum known depth in the crust of 10,500 feet. They originated in the late Cretaceous or early Tertiary periods 60 to 70 million years ago.

Both in the basin's northeastern and southeastern quadrants, a few of the great fissures have been named, and are generally designated as "fracture zones." Thus, from north to south, we have the Chinook, Mendocino, Murray, Molokai, Clarion, Clipperton, and Galapagos in the first-named quadrant; the Easter Island (Fracture Zone), Challenger, Fernandez, and Eltanin in the second.

The Mendocino Fracture Zone (sometimes called the Mendocino Escarpment) would appear to be almost in a class by itself. Named for California's Cape Mendocino, which it almost touches at the famed San Andreas Fault, it extends westward into the Pacific for 1,500 miles, is 60 miles across at its widest point, and in its central area penetrates the crust to a depth of 19,500 feet below sea level. The first Pacific fracture zone ever subjected to magnetic tests revealed an 80-mile lateral shift across its fault line, an impressive but not too uncommon occurrence in this region. The most recent study of the Mendocino Escarpment, however, is another matter. Relative to its south side (which is moving easterly), its north side has been displaced 750 *miles* to the west. What are the colossal forces capable of bringing about such large-scale movement in the crust? And what will occur if such a moving fracture splits a continental margin? Scientists can only speculate.

One might expect that the deepest depressions of the sea floor would be found near the centers of the deep-ocean basins, but this is not the case. These *trenches,* as they are called (the Pacific has more of them than any other ocean), lie quite near the continents and in fact closely parallel both the earthquake zones and the successive lines of volcanoes. To name just a few, at the basin's northern end there lies the Aleutian Trench, an enormous elliptical gash in the earth's surface running all the way from the narrow shelf of the Kamchatka Peninsula to the Gulf of Alaska, its span of nearly two thousand miles making it the second longest on earth. Moving south, there is the Middle America Trench, a depression stemming from the Gulf of California, extending southeastward to the Andean bulge below Panama. Then there is the Peru-Chile Trench, the longest in the world. Running the full length of South America's west coast, it terminates south of Cape Horn, roughly along that 70th meridian which may be said to divide the Pacific from the Atlantic.

In the western reaches of the basin the picture is a bit different. Here, instead of lying close to continental shorelines (and conforming to their shapes), the trenches flank various island arcs, usually on their seaward sides rather than between them and the major land masses. An intricate system of trenches lies north and northwest of New Zealand, including the Kermadec, Tonga, Vityaz, New Hebrides, and New Britain depressions. Leaping New Guinea, moving past Australia, still another system branches northwestward, the Great Java Trench (another ellipse like the Aleutian) bordering the islands of Indonesia. Were we to sail north from eastern New Guinea, our keel would furrow the water above the Mariana Trench (the Philippine and Ryukyu trenches would lie to the west), and beneath us would yawn the mighty Challenger Deep. Still farther north, the Izu, Japan, and Kuril trenches would form one unbroken line all the way to the western Aleutians.

Both as to shape and depth the trenches are incredibly uniform, describing a narrow V in cross section, the majority of the deeper ones "bottoming out" at about 35,000 feet below sea level. Many of the more shallow are still obviously in the throes of creation; others are old, and over the centuries have accumulated sediment. Still, geologists believe their average depth to be the same, about 35,000 feet.

Without a doubt, the trenches are one of the greatest of geological mysteries and many questions about them have yet to be answered. Obviously, it is more than coincidental that they so closely parallel known lines of volcanic and seismic activity. Obviously, too, a pattern exists, the various facets of which are bound to be revealed as marine geologists continue their exhaustive studies: (1) Why is it that gravitational attraction in the area of the trenches is the lowest to be found on earth? (2) Why are heat-flow measurements in these zones only about half that of the crust's average? (3) Why are quakes originating along the trenches always of shallow focus, while those occurring landward of them show a progressive increase in depth? (4) What internal processes have caused the trenches to develop?

Speculating on the significantly reduced heat flow within the trenches, some scientists have theorized that it may be due to convection currents in the viscous fluid of the mantle. You may observe such currents at work in a kettle of boiling tomato soup. Moving away from the source of heat, they rise, reach the surface, then travel to the outer edge, sinking again as they cool. Thus it may be inside the mantle. If so, it is the trenches which mark the zones where comparatively cool rock is moving downward, tending to fracture the crust at those points, dragging it down beneath the lighter sial of the overriding continents. Certain tangible factors lend support to this theory. For one thing, the material of the continents and the sea bottom, being lighter than that of the mantle, should theoretically float on the latter at every point, remaining always in a state of equilibrium (called by geologists "isostatic equilibrium"). That it does not do this in the trench zones, but instead is being sucked down into the mantle, means that a "negative anomaly" is present there; i.e., that the natural buoyancy of the crust is being negated by the downward pull of the forces creating the trenches.

Another possible indication supporting the theory of "convection currents" in the mantle is derived from heat probes. It has been noted that certain zones within the crust are areas of comparatively high heat-flow. During the IGY investigations (International Geophysical Year, 1957–1958), such conditions were discovered to exist near the Easter Island Rise off South America's southwest coast where the heat passing through the ocean floor is *seven times* that of the crust's average. Here, molten rock may be welling up from the mantle at one of the high points of the convection cycle while in other regions it is sinking. This would imply the simultaneous existence of more than one current in the mantle. Geologists subscribing to the theory refer to these separately circulating swirls of fluid matter as "convection cells."

As has been indicated, the canyons carved in the slopes, the trenches, fracture zones, and abyssal plains are not the only fascinating features of the Pacific Basin. There are many mountains (mostly apparently volcanic in origin); some with pointed summits called *seamounts;* others, called *guyots,* flat-topped and of an amazing similarity of shape. There are many coral atolls, a study in themselves and equally intriguing. There are also sharp and jagged ranges, transversely split by the fracture zones mentioned above, variously called *ridges* or *rises.* To the latter, as to other features of crustal instability, our examination must in general be confined.

The most prominent of these submerged ranges is the East Pacific Rise, an enormous bulge in the sea floor that begins off southern Mexico, runs the full length of South America (mostly between the 105th and 120th meridians), swings past Antarctica, flanks Australia on the south, and merges with the Mid-Oceanic Ridge in the Indian Ocean. Only in two places does it break the sea's surface (the positions marked by the Galapagos and Easter Islands), but, gigantic as it is, the East Pacific Rise is but one link in a chain. The Mid-Oceanic Ridge joins the Mid-Atlantic Ridge, itself 10,000 miles long. Taken together, the three ranges dwarf all other earthly features; a 40,000-mile-long submarine cordillera.

Certainly, one does not have to be a geophysicist to determine that the earth's crust is in a constant state of change. That much is evident. What is not so easily understood is exactly how it is changing and why. In 1910, the German meteorologist, Alfred Wegener, was a member of an expedition making a study of Greenland, that vast, incredibly desolate island whose icecap covers 700,000 square miles and reaches in places a thickness of 11,000 feet. As the men worked, they had earlier surveys in mind. For example, it was known that in 1823 Sir Edward Sabine had discovered and plotted the small island off the east coast which later bore his name. There was nothing unusual about that. However, when a second party of geographers again plotted the island's position in 1869, their calculations placed it a *quarter of mile west* of where Sabine had marked it on his map. Wegener could not understand this. In both cases the men were professionals, extremely precise in their figuring, which had been repeatedly checked. Was it possible, he wondered, that *both* sets of figures might be accurate, each for its given time? Could Sabine Island's location have *shifted* during those 46 years? Taking new measurements, Wegener disovered an exciting fact. *The island had moved again.* It now lay ⅝ mile farther to the west than it had in 1869.

Finding this phenomenon extremely intriguing, Wegener felt compelled to delve further into the mystery. Using the latest scientific methods developed up to that time, he determined the positions of other land masses of the Arctic, concluding that the previously detected westward drift was common to them all, though the rate of movement varied. The Americans, he believed, were annually widening the gulf between themselves and the Eurasian-African continental complex by a distance of $1/_{25}$ inch. Within a few years he thought he had gathered sufficient data to formulate a tentative hypothesis. The result was the *theory of continental drift*.

According to this theory, at approximately the opening of the Lower Carboniferous Period of the Paleozoic Era (330 to 345 million years ago), the continents were congealed into a single land mass (called Pangaea by Wegener) surrounded by a single world ocean (which he named Panthalassa). During the passage of millions of years this "supercontinent" split into several large segments, each of which drifted very slowly apart from the others. North, Central, and South America moved westward. Australia broke away from India and slid southeastward across the Indian Ocean.

Though this idea may appear preposterous at first thought, a careful study of a map of the world will make it seem less so. Clearly, between the Caribbean Basin and the east coast of South America on the one hand, and the west coast of Africa on the other, there exists the possiblity of a rough fit. So too, between eastern North America and western Europe. When to this factor the following considerations are added, the theory more closely approaches credibility: (1) Comparison of the dating of West African rock formations with those on the Brazilian coast place them in the same time period, as do those of South Africa and Patagonia, western Scandinavia and Canada. (2) There appears to be a definite relationship between certain of the flora and fauna of both regions. (3) Australia, India, South Africa, and part of South America reveal topographical evidence of prehistoric ice sheets and glaciers. (4) Deposits of coal in North America, Europe, and Antarctica could mean they once spanned the Equator.

Wegener's thesis, however, though impressive in some of its aspects, was challenged almost from its inception. Scientists argued back and forth. In 1926, the International Astronomical Union entered the debate and began taking measurements in North America. The results hardly clarified anything. By 1935 they were able to announce that Vancouver and Ottawa were closer by 19 feet than they had been nine years earlier, but that during that same time, San Diego, California, and Washington, D.C., had moved farther apart by 40 feet. Their basic contention was the reverse of Wegener's. According to them, the American and European continents were not separating but very gradually closing on each other.

Today, while the theory of continental drift has not been abandoned, it has undergone substantial modification. Although there are numerous hypotheses (and new ones are constantly being formed), only a relative few continue to hold the focal point of attention. The *contraction theory* of continent building holds that a gradually cooling mantle induced a shrinking and cracking of the crust, that the oceans were formed by escaping steam, the atmosphere by gases, the continents by extruded lava (magma expelled through surface fissures). Partially, the crust shrank because of surface cooling, partially because of the drainage from beneath of extruded products. As shrinking increased, so did fissuring. Through the resultant creation of these *zones of weakness,* mountain ranges rose.

The *convection current theory* has already been mentioned. Among its adherents are Maurice Ewing and Bruce Heezen, two of the

most distinguished marine geologists at Columbia University's Lamont Geological Observatory. Such men would probably take exception to the figures released by the International Astronomical
Union. They see the flow within the mantle and the consequent
crustal movement as an infinitesimally slow process, possibly not
more than an inch annually. This is extremely small by calendar
measurements. However, when placed in the immense perspective
of geologic time it becomes significant. An inch a year works out
to one mile every 63,360 years. According to such reasoning, over
a span of 63,360,000 years, a given land mass subject to this rate
of movement would cover a distance of *1,000 miles.*

Theoretically, there are three well-defined stages in the process
of continental drift. The first, the fracturing of a major land mass,
is exemplified by the Great Rift Valley of East Africa, an enormous,
complex fissue, interrupted in the middle by Lake Victoria, running
from the Gulf of Aden in the north to the Zambezi River in the
south. This rift is geologically young and has thus far split but once.

As an example of the second stage, the Red Sea could be cited.
This is an old rift and fairly broad. Underwater maps of the area
show a much narrower and newer rift threading its center for nearly
its full length.

The classic example of the third stage is the Atlantic Ocean, a
split now three thousand miles wide. According to the theory, it
too was once as narrow as the present African fissure, and in the
dim past divided the supercontinent that existed before the Amer-
icas broke apart from Africa and Europe. Within this great basin
another, much newer rift has appeared, the recently discovered
Mid-Atlantic Ridge.

To these two basic hypotheses we may add a third, actually an
adjunct to the idea of convection currents. Opposed to the theory
of contraction, Dr. Heezen believes that the earth is actually *expanding,* opening up over a multimillion-year-long process like an
overripe fruit. He cites the 40,000-mile-long Mid-Oceanic Ridge
girdling the planet and the giant crack that threads its entire length.
Is this position tenable? No one can positively say, but one thing
seems certain. Regardless of what may be occurring within the
overall process of continent-building, in given areas the planet's
crust *is* splitting apart. The process is slow, very slow, but it is also
very sure.

*   *   *

Perhaps, by now, the reader has at least a rudimentary idea of the structure of the earth, or, more accurately, of the basic concepts behind current theory. Our subject is seismic sea waves. However, since such waves are generated as the result of natural occurrences within the earth's crust (or just beneath it)—and since, morever, such crustal movements are frequently affected by the mantle, and the mantle by the core—it is felt that some understanding of this structure is essential for the study that is to follow.

3

Undersea Volcanic Eruptions and Submarine Earthquakes

Krakatoa, "The Blast Heard Round the World!"—and the waves it created. Lieutenant Billings's account of the end of Arica. Tsunami on the open sea.

It was shortly before 10:00 in the morning, a Monday, August 27, 1883. From high on the shoulder of the mountain he was climbing, the Dutch vulcanologist Van Gestle looked down through an atmosphere thick with volcanic ash, hoping for another of those periodic moments of lesser murk that would enable him to view the sea. He had climbed before in these hills. Never had he climbed on a day like this. Below him, he knew, on the far side of the dense, intervening jungle, lay the coastal town of Anjer Lor, and a few miles to the north of it the modest, part Chinese settlement of Merak. Just now he couldn't see either of them,. For that matter, he couldn't see any part of the coast of Java, not a yard of the waters of the Sunda Strait, and certainly not that ring of volcanic islands which the Batavian Government had sent him here to observe.

He closed his eyes a few moments to sharpen the image, then, squinting, attempted to refocus, as though by some miracle his vision might penetrate the haze. Even at this distance of 30 miles he could hear the ominous rumbling of Krakatoa, and every moment he expected another great blast such as the two that had rocked the coast during the pre-dawn hours.

He raised his binoculars, leveling them in the general direction of the triangle of volcanic islands, seeing them again in his mind's eye. Two of them, Lang and Verlaten, were very small, and closely flanked Krakatoa to the northeast and northwest, respectively. Krakatoa itself was much larger and had three peaks, Rakata, Danan, and Perboewatan. Rakata was an almost perfect cinder-cone, 2,623 feet in height, or had been before this morning. No one knew what it looked like now or whether it still existed. Only its violent eruptive state was a certainty. Since its last eruption in 1680 the volcano had remained quiescent, in fact, had been considered extinct. But only three years ago there had been an earthquake, its epicenter close to the mountain's base. And this year, in May, it had again become active.

Searching the invisible sea for a break in the all-obscuring pall of ash, Van Gestle once more became conscious of the handkerchief tied over his nose and mouth. It was saturated by now and of little use. The burning of his skin and the increasingly bitter taste on his tongue told him the sulfur dioxide concentrations were rapidly increasing, combining with moisture to load up the atmosphere with sulfuric acid. At this point it was sufficiently dilute to permit human survival but it might not remain so much longer.

He searched. He scanned the murk. He continued to see nothing. Again, he put on the goggles he'd worn earlier, hoping to protect his smarting eyes. As before, the lenses glazed over immediately, the accumulating film of ash forcing him constantly to wipe them. He thought about other East Indian eruptions, the recent and the not so recent. Papandayang on Java; 1772. She'd blown herself to pieces, losing 4,000 feet in height. Tambora on Sumbawa Island; 1815. The Englishman, Sir Stamford Raffles, had written about that, left a vivid account. Enormous blast. More than 4,000 feet of summit pulverized, 36 cubic miles of matter. Twelve thousand killed, four times the casualties of Papandayang. They'd heard Tambora in Sumatra, at points 1,000 miles distant. The skies over Java had turned boiling black. Great whirlwinds had risen, sucking men and animals into the inferno. The trees that were felled by the thousand had all pointed *toward* the crater.

He shifted the binoculars. As a vulcanologist he'd come to the right region, he thought, trying to control his excitement. No area on earth had as high a degree of vulcanicity as this. Why, only last year Java's Galunggung had erupted, claiming more than a hundred villages and 4,000 lives. Only last year. . . .

Suddenly, as though a fierce wind had risen and was blowing

over the sea, the ash-choked atmosphere seemed to clear itself. Van Gestle tensed. In an instinctive gesture almost of frenzy he ripped the goggles from his eyes and again raised his glasses. Now he could see it as by a miracle. Spread before his gaze was the wildest scene he'd ever witnessed, weird, like something out of a nightmare. The ocean had gone insane. Its waters had no discernible pattern but seemed to be churning and thrashing in every direction at once as though stirred from underneath by some gigantic hand. And there was Krakatoa right in the middle of it all, smoking more than ever, glowing bright red now at the crown of its sharply pointed main peak. There were boats in the bay beneath it, all of them abandoned days ago by crews who feared the swamping of small craft in such a sea.

And now, as he watched, the impossible happened. As though drawn by some irresistible magnetic force, the boats were moving across the bay, all very swiftly and in the same direction with no human hand at the helm of any of them. In seconds they vanished, engulfed in a fearful maw of steam and roiling ocean. Directly across the strait an arrow of yellow fire pointed toward the island, marking the line where the earth's crust had opened in a vast fissure, where uncounted trillions of gallons of cold water were now penetrating to mix catastrophically with the white-hot magma seething beneath. Enormous clouds of superheated steam burst upward, rising nearly instantly to thousands of feet in height. Van Gestle held his breath. It was exactly 10:00.

At 10:02 there came one of the greatest explosions ever heard by man, a thunderous, earth-shattering detonation with a force equivalent to a least 100 megatons, a blast that utterly obliterated 5 square miles of the planet's surface and literally atomized most of what remained of the Krakatoa Islands, leaving a crater 3½ miles across. It was heard in the Philippines, in Australia and New Zealand, even on the island of Madagascar at a distance of almost 3,000 miles. Van Gestle was knocked flat by the shock wave. When he regained consciousness he could see nothing at all. Though it was still morning, an impenetrable blackness had descended over Java.

Various writers have pointed out that, since the great waves which followed were the result of a volcanic eruption rather than crustal faulting, Krakatoa was very unusual in modern history, tsunami being rarely due to such a cause. This is only partly correct. It is true that no waves of comparable power had been generated by volcanoes for nearly 33 centuries, but what is sometimes overlooked is the fact that Krakatoa itself lay at the intersection of two converging fault lines of enormous extent. Both of these colossal fissures ran completely across the Sunda Strait in a vast X-shaped configuration, meeting each other at a point immediately adjacent to Krakatoa Island's north coast. Striking the shorelines of both Java and Sumatra, penetrating inland for unknown distances, they created then (and, presumably, would still retain their deadly potential today) a zone of very violent seismic activity, and indeed stood at the center of the most volcanically unstable region on earth, the Indonesian Islands.

In his writings, Van Gestle makes it clear that he saw but a single line of flame stretching across the strait, an indication that only one of the two great faults was sufficiently deep to connect with the volcano's magma chamber or a subterranean conduit leading to it. But which one? Van Gestle was climbing a mountain above the settlement of Anjer Lor. Hence, Krakatoa was nearly due west of him. He saw the boats in the bay, saw them sucked down into this fiery cauldron. The streak of flame he viewed, therefore, could only have risen from the fault which crossed the strait from the southeast to the northwest, since between this bay and the other fissure the island of Krakatoa would have intervened. It was this great crack, running from a point on the Java coast a little southwest of Anjer Lor all the way to the Semangka Peninsula, which in all likelihood played as prominent a role in the creation of seismic sea waves as did the mighty eruption itself.

The first signs that Krakatoa was again awakening to life were evidenced in May, when warning rumblings from deep within the larger cones of Rakata and Danan rapidly built to thunderous blasts audible in Batavia (today called Djakarta), and Buitenzorg (today Bogor), 100 miles distant. The Batavian government took these signs seriously. Pleasure trips to Krakatoa by steamer were canceled. Throughout June and July the mountain continued to erupt, filling the atmosphere with sulfur fumes that fouled the air over the strait. Volcanic ash fell 300 miles away. New vents opened in Rakata. Though the intensity of the eruption fluctuated, activity was continuous. On August 17, nearly two weeks before the culminating explosions, in the log of the bark *West Australian* was noted a singular entry. According to her skipper, Captain Thomas, his ship had plowed through great masses of floating pumice for 500 miles, and now, after three days, had broken through it. At the time he was in the Indian Ocean, 1,000 miles from Krakatoa.

Early on Sunday evening, August 26, like a warning given by fate itself, the signs of impending disaster increased with the appearance of the first small waves. Characteristically, they spread out in concentric rings. One struck the town of Teluk Betung on Sumatra's Lampong Bay. Another hit Tyringin on the Java coast north of Pepper Bay. The one that passed over the Chinese camp at Merak was only 5 to 6 feet in height. All caused some deaths and moderate damage. Because no precise record of the arrival times is now obtainable, it is impossible to say whether these were different tsunami or simply widely separated sections of arc within the same wave. But one thing is certain. Since they occurred *prior* to the great blasts of the following morning, they were caused by crustal slippage, apparent proof that at least one of the two huge cracks existed before the final cataclysm.

Volcanic activity increased during the night, the rolling reverberations disturbing the rest of people as far away as Batavia. Before dawn on Monday morning the first great explosion occurred, marking the distintegration of the 400-foot peak of Perboewatan. This part of the island simply sank. Into the hole where it had stood the sea poured.

In less than half an hour came a second mighty blast. According to the reports of geologists who later reconstructed the scene, this must have been the moment when the 1,500-foot Danan peak disappeared, allowing additional millions of tons of seawater to rush below into the white-hot molten depths. The steam pressure thus created, now, with enormous rapidity, began to build the incalculable energy which within a few hours would produce the greatest explosion of modern times.

As concerns their effect on the sea, these two pre-dawn blasts produced immediate and catastrophic results. Each generated a huge tsunami. Striking during a night whose darkness was intensified by heavy ash fallout from the mountain, the tremendous impact of these waves was taken by the coasts of Java and Sumatra.

Teluk Betung and Anjer Lor were inundated. Thousands drowned. The number of survivors is unknown.

About the third wave there is controversy to this day. Depending on the texts one consults, estimates of its height vary considerably, but since the same wave will normally rise to different heights in different localities (depending on the shelf topography at a given point), it is possible that nearly all the estimates may be more or less correct. This wave followed the third and mightiest explosion, the one coming at 10:02 A.M., and completed the work of the first two. To give some idea of its awesome power, one might reflect on the tsunami that struck the city of Hilo, Hawaii, in May of 1960. On that occasion, there were several waves. Gene Wilhelm, a reporter for the *Hilo Tribune-Herald*, wrote of one of the largest of them as follows:

Whole buildings were driven directly through the ones behind, the way you'd close a telescope. Boulders weighing nearly a ton were picked up and set down hundreds of yards away. And through it all there was the sickening thunder of buildings disintegrating, with the screams of people trapped inside.

Hilo (since rebuilt) was a modern city. The buildings referred to were of steel-reinforced concrete. Yet waterfront photos taken after the catastrophe present a scene of desolation reminiscent of the bombed-out cities of World War II. How big was the wave which drove whole buildings "directly through the ones behind"? *Its height was 35 feet!* How then can we estimate the pulverizing force of a wave which, according to the most conservative reports, rose in at least one locality to *four times* that height? Obviously, we cannot. We can review the casualty list of nearly 37,000 dead (close to *five times* that of the 1976 Philippine tsunami), almost all of them either drowned or crushed by the waves. We can try to understand. But in the final analysis it is just a figure, very difficult to comprehend in terms of human life.

Logically, we can assume that many of the victims were killed by the first two tsunami, those which struck before dawn and annihilated a multitude in their sleep. But a good many more must have fallen to the third wave. There was no one to tell them about tsunami, a phenomenon preserved on both islands in tribal legend, but about which there was no definite knowledge, no living person old enough to remember the last one. They could not have anticipated the third volcanic explosion. It is probable that hundreds of evacuees (perhaps thousands), returning to the debris of their towns to search for lost loved ones, did so only to meet their doom. In the dense murk of heavy ash fallout they couldn't have seen the third wave. Krakatoa's final blast of self-destruction would have come, its shock-wave of fantastically compressed air knocking people to the ground. Minutes later, invisible as it rose out of the sea, the enormous tsunami would have struck.

The statistics are revealing. Merak, which had received only moderate wave damage the night before, was now to perish. The bay on which it faced was funnel-shaped, with the town at its narrow inland end. When it entered this bay, the third and greatest tsunami was already an estimated 50 feet in height, its waters rising as contact with the shoaling sea bottom slowed them from over 400 miles per hour to perhaps half that. But the bay's constricting headlands slowed it still more, forcing the trillions of gallons of fast-racing water behind it to pile higher and higher. By the time it curled its incredible bulk above Merak its speed was down to less than 100 mph, perhaps as low as 50. But its height was an estimated *135 feet!* Against this towering wall of death no man-made structure or living creature could stand for an instant. The sheer power of its impact would have been nearly incalculable.

So died Merak. Anjer Lor was now struck again, this time by a 33-foot wave, while Tyringin, twenty-four miles from Krakatoa, was smashed to kindling by a 70-foot tsunami.

Across the Sunda Strait, Sumatra was equally hard hit. The town of Katimburg was only nineteen miles from Krakatoa. An 80-foot wave utterly crushed it. Teluk Betung, already overridden in the pre-dawn inundation, was engulfed again. Government House was on a hill behind the settlement 78 feet above sea level. Surging water came within six feet of the structure, drowning the already stricken town and its remaining inhabitants. And it did something else. In the harbor the Dutch warship *Berouw* had been anchored. The wave swept it up, carried it completely over the shattered city and 1.8 miles up on the course of a partially cultivated valley. There, largely undamaged, the ship was left at 30 feet above sea level where her rusting hulk may still be seen today.

There were other effects. Everywhere between the main islands, lighthouses and their crews were destroyed. Twenty minutes after the third explosion pulverized Rakata's cone the harbor of Batavia and all its ships and boats were obliterated, with shipping losses throughout the strait totaling an estimated 5,000 to 6,000 vessels and small craft. In addition to the major towns, 295 lesser settlements were inundated, and over a thousand native villages, with the appalling loss of life already noted.

The waves spread across the ocean. Five hundred miles from Krakatoa, tide gauges in Surabaya Strait were driven 10 inches up the scale as the tsunami passed. Nine hours after the culminating blast, 300 riverboats were swamped and sunk at Calcutta in the Ganges Delta over 2,000 miles away, while in Port Elizabeth, South Africa, 5,000 miles distant, large ships strained at their anchors and threatened to part their cables. Even in the English Channel, 11,000 miles from Krakatoa and halfway around the earth, the waves were felt, recorded by tide gauges there some 32 hours after the eruption.

Much more might be said. One might speak of the sound waves from the blast, how the police station 3,000 miles away on Rodriguez Island recorded: "The distant roar of heavy guns, coming from the eastward." Or of the fact that instruments in Europe and on both coasts of the United States recorded significant changes in barometric pressure on nine successive days following the eruption. Or of the weird but gorgeous atmospheric effects caused by worldwide jet-stream dissemination of enormous quantities of volcanic dust. Ten years after the event this dust was still detectable in the troposphere.

Such facts as these enthrall the mind of man and quicken the scientific imagination. Nonetheless, when all the talk of Krakatoa is done and all the statistics evaluated, one salient fact remains. Despite the frightful power of the eruption itself, the emission of noxious gases, flame, smoke, volcanic bombs, and other ejecta, it was not the mountain's fury which destroyed those unfortunate thousands, except indirectly. It was the terrible power of water. In the vast majority of instances death came at the hands of seismic sea waves.

*   *   *

Exactly what happens in the earth's crust to cause tsunami to develop, how do they form, and, once generated, what is the nature of their oceanic movement and terminal modification on arrival at a coast?

Initially, such waves are usually created not by volcanic eruptions but by the faulting and slippage of a part of Earth's crust due to a submarine earthquake. Normally, this slippage is vertical, though on occasion it may be lateral or may even embrace a combination of the two movements. When the quake strikes, causing a rising or subsidence of the sea floor, the effect is for the sea to equalize the resultant sudden change in pressure by duplicating this configuration at the surface. If the sea floor subsides, the water at the surface will become similarly depressed and a vast "hole" will form. If it rises, the surface becomes domed, sometimes to a great height. In either event a profound oscillation is generated (in the case of a major slippage it may be of titanic magnitude), and the tsunami move away from their point of origin in ever-expanding concentric circles, one after another. On occasion, there may be as many as five or six successive waves, and, rarely, as many as nine.

Though the wave may be very high at its source, it is quickly flattened by gravity, and in some respects resembles a normal longitudinal wave. However, its wavelength (distance from its crest to that of the wave that succeeds it) may be enormous, as much as 600 miles in the broad reaches of the Pacific (though far shorter in the narrow Mediterranean), while its height from crest to trough may be 2 feet or less on the open sea. Accordingly, it proceeds as a wave of the shallow water type, its speed at any given point proportional to the square root of the depth of water over which it is passing.

The velocity of tsunami is readily computed by the following formula:

$$v = \sqrt{gD}$$

In this simple equation, v is the velocity of the wave, g the gravitational acceleration, and D the depth of the water. For example, the average depth of the Pacific is approximately 3.5 miles or 18,480 feet. With the value of g as 32 feet per second per second we arrive at the following:

$$v = \sqrt{32 \times 18{,}480} = \sqrt{591{,}360} = 768.9993 \text{ feet per second}$$
$$= 769 \text{ feet per second} = 524 \text{ mph}$$

Admittedly, this speed is impressive, but over certain parts of the sea such waves will travel much faster. For example, a tsunami traversing an area having 30,000 feet of depth will move at 670 mph, as fast as many jet aircraft. It has already been indicated that the wavelength is the distance in miles between two successive crests. This too can be computed, but to do so one must know both the velocity of the tsunami and its period, i.e., the length of time required for two succeeding crests to pass a given point. As an example, we could take the tsunami generated by the great Chilean earthquake of 1960, a disturbance originating under the sea in the Peru-Chile Trench. To travel the 6,600 miles across the Pacific from the Chilean coast to Oahu, this wave required 14 hours and 56 minutes, averaging 442 mph. To cover the much greater distance of 10,600 miles to Honshu, Japan, the changing sea floor depths of the Pacific caused the same wave to average the slightly greater speed of 480 mph. It had a period of approximately 60 minutes, a fact determined by widely distributed tide gauge stations throughout the Pacific Basin. The formula is as follows:

$$\lambda = vt$$

Here, where v is the tsunami's velocity and t is its period, substituting into the formula, we compute the wavelength, λ, at approximately 520 miles.

As will be easily seen, waves the distance between whose crests is anywhere from 60 to 600 miles, and whose height on the open sea is usually not over 2 feet, are not going to be detectable by shipping. Despite the tsunami's great speed, therefore, it will pass under a large vessel without being noticed and without effect. It is only when it reaches a coast that its quiescent phase ends and it undergoes its terrible and often tragic metamorphosis. Here, where the continental shelf rises, this source of incalculable energy, a foot or two in height (but with hundreds of miles of breadth in any significant portion of its arc) encounters shoal water for the first time. The friction-caused drag effect of the shallow bottom, becoming more shallow with each second of the wave's progress, is to slow it at once and slow it drastically. Thus, over a mile or two of distance on the shelf, a 500-mph tsunami may have its speed cut to half that, then to a quarter and even less. But meanwhile, as the wave front slows, the water behind it begins to *pile up,* millions of tons of it rising with fantastic rapidity to greater and greater heights. If the stretch of coastline thus struck is more or less straight and unbroken, the wave's height may be no more than 30 or 40 feet. But if it is an *embayed* coast—and if, in addition, the gradient of the slope at a given point is extreme—the portion of the wave impacting in that area becomes constricted and its energy concentrated. It may then rise to great heights and in a populated region wreak untold havoc. As one example, in the year 1737, there was the tsunami which struck Cape Lopatka on the southern end of what is today the Soviet Union's Kamchatka Peninsula. Fortunately, the population was sparse. The wave's height was recorded at 210 feet, as high as a 20-story building!

It should now begin to be clear why a 45-foot wave of the normal type may be ridden at Waimea Bay by surfers who survive to tell of it, while a 6-foot tsunami (such as that which first struck Merak on the day prior to Krakatoa's culminating blasts) took its toll of lives. The waves at Waimea, though huge in size, have relatively little behind them. Although they too may come from thousands of miles away, they have a short wavelength and a brief period. Because they are the result of wind and tide rather than crustal dislocation; because, moreover, their deceleration on meeting shoal water is in the order of 60 down to 40 mph, rather than 400 (or even 700) mph down to less than 25 percent of these velocities, the sea does not pile up behind the Waimea waves, and each of them is a separate, unsupported entity unto itself.

Usually, but not always, the earliest indication that a seismic sea wave is on the way is seen when the tide goes out very abruptly and to a degree never witnessed under normal circumstances. This occurs when a submarine earthquake has caused crustal fracturing, when the resultant collapse of the sea floor creates a vast depression on the ocean's surface. Immediately, the sea restores this disturbed equilibrium by sweeping in from all sides to close the depression, sucking countless tons of water away from the coastal areas at a rate far in excess of the lowest tidal ebb. This enormous surge flowing rapidly seaward creates its own imbalance, overcompensating in the area of the disturbance, not only filling the depression but tending vastly to overfill it. The result is a colossal countersurge again moving landward, a sort of backlash effect. This is the tsunami.

Following the initial suboceanic shock, this phenomenon may repeat itself several times as the crustal fractures multiply or a single fault continues to rise and subside. Although (if the quake's epicenter is close offshore) the first wave will usually strike within fifteen minutes of the time of the disturbance, it may not come for much longer than that, depending on the distance factor. Thereafter, successive waves may arrive with periods varying anywhere from ten minutes to an hour, and in some cases even longer. Because of this time lag people have lost their lives. In some instances, not knowing that other waves normally succeed the first one, they've returned too soon to a stricken area only to be caught and destroyed. In other cases, no wave having as yet arrived, they've actually *waited* for it.

This happened at Hilo, Hawaii, on May 23, 1960. The city inhabitants had been warned. They knew a tsunami was coming. They might have remembered the 1946 wave, 50 feet high, a killer that took 159 lives and claimed $25 million in property damage. Instead, a number of persons went down to the waterfront *to watch the wave come in.* The result was predictably tragic. So far as is known, the "sightseers" were wiped out to a man. Within the city itself, several square miles of the business district were smashed flat, raising the death count to 61. Afterward, Kamehameha Avenue had the typical "blitzed" look, only scattered rubble remaining.

It has been noted that on some occasions there is no sudden and dramatic lowering of the tide, no natural warning whatever. This is the case when there is no depression in the sea floor but rather an upthrusting, the rising of a vast mound that occurs without linear fissuring, without an extensive fracture through which the water can flow downward. This is a much less frequent occurrence but it does occasionally happen. Normally, it is the result of submarine volcanic activity, especially at the birth of a new volcano or one that has risen from the sea bottom by successive eruptive stages until its summit finally emerges into the air. Shortly before such emergence the sea's surface becomes extremely agitated. The water boils. Fish life dies. Steam escapes from the depths in ever-increasing quantities. Black jets of ash rocket skyward, often forming thick clouds that can acquire a characteristic "pine-tree" configuration. Smothered by the ocean's weight and density, the eruption progresses in an eerie, "unnatural" silence. Only when the ravaged summit lifts above the waves does it become audible. Only when there is a steam-caused explosion of cataclysmic intensity does it produce a tsunami.

Volcanic islands are created fairly frequently. Many have short lives, their uncongealed lava and porous ash quickly eroded by wind and wave or destroyed by a continuation of the explosive processes which created them. Although they are seldom the cause of seismic sea waves, it is entirely probable that a certain percentage of them do cause such waves. The newly born island would disintegrate in blast and fire. Tsunami would be generated. If the location were far at sea, no seismograph would detect an epicenter. Why not? Because, unlike the great tectonic shocks (generated by earthquakes) whose depth of focus is nearly always shallow, volcanic temblors originate at such depths that not many are detectable over any radius exceeding a few miles. Needless to say, if the detonation in which the island dies is great enough, instruments *will* detect it, as was the case with Krakatoa. Here, even if primary and secondary waves had not been propagated within the body of the Earth, the tremendous concussion still produced air oscillations resulting in worldwide fluctuations in barometric pressure. Microbarometers did not exist in 1883 but they weren't needed for that. Ordinary barometers were sufficiently sensitive to do the job.

Already related was the memorable "voyage" of the *Berouw*, the Dutch warship which was carried by the third wave from Krakatoa completely over the drowned and shattered city of Teluk Betung. Such dramatic incidents have occurred more often than would be supposed. Another example is that of the U.S.S. *Wateree* in 1868, a single-stacked, two-masted side-wheeler which on August 8 of that year was anchored in the previously twice-destroyed Chilean port of Arica. It is doubtful that her captain had given much thought to the region's history of seismic sea waves. He may never have heard of the tsunami which destroyed two Spanish galleons at Valdivia in 1575 or of the one that leveled Concepción in 1751. But he would learn. Moored near his own in Iquique Bay that day were several other vessels, including a swift Peruvian warship and a merchantman of American registry. To Lieutenant L. G. Billings of the *Wateree* we are indebted for one of the most graphic eye-witness accounts of a tsunami ever recorded:

I was sitting in the cabin with the captain towards four o'clock in the afternoon when we gave a sudden start: the ship was trembling with the same vibration that occurs when the anchor is let go and the chains thunder in the hawse-holes. We knew that it could not be that, and we ran on deck. ur eyes were at once caught by a huge cloud of dust over the land, which was coming up from the southeast while the terrible thundering grew louder and louder. As we watched, stupified, the hills seemed to be capsizing, and the ground moved like the short choppy waves of a rough sea.

The cloud swallowed up Arica. In that very instant, through its impenetrable veil, there arose shrieks for help, the din of falling houses, and the thousand mixed noises of a great calamity. Meanwhile, our ship was shaken as if by the grasp of a gigantic hand. Then the cloud passed on.

As the dust thinned out we rubbed our eyes and stared, unable to believe what we saw: where a few seconds before there had stood a happy and prosperous city, busy, active, and full of life, we saw nothing but ruins. The less seriously wounded of the unhappy people caught under the wreckage of what had been their houses were struggling among the ruins, and everywhere shrieks, cries of pain, and calls for help tore the air under the pitiless sun shining in a cloudless sky.

We were worried about the coming of a tsunami, and we put out to sea; but the water was calm and unruffled, and it might have been supposed that the four or five minutes that we had just passed through as well as the shockingly distressing scene upon which for the moment we were turning our backs were part of a nightmare. Nevertheless, as a measure of prudence the captain set out extra anchors, had the hatches closed, the guns lashed and lifelines rigged.

Meanwhile on shore the survivors were coming down the beach and crowding on to the little jetty, calling to the crews of the ships to come and help them get their relatives out of the twisted ruins and carry them to the apparent safety of the anchored vessels. This was more than we could withstand, and the yawl, with thirteen men aboard, was launched at once. It reached the shore and its crew got out, leaving only one sailor to guard the boat. We on board were in the act of organizing a body of forty men to be sent ashore with axes, picks, and shovels when all at once a hoarse murmuring noise made us look up; looking towards the land we saw, to our horror, that where a moment before there had been the jetty, all black with human beings, there was nothing: everything had been swallowed in a moment by the sudden rising of the sea, which the ship, floating upon it, had not noticed. At the same time we saw the yawl and its sailor carried away by the irresistible wave towards the lofty, vertical cliff of the Morro, where they disappeared in the foam as the wave broke against the rock.

At that very instant there was another earthquake shock, accompanied by a terrible roaring on the land that went on for several minutes. Once more we saw the ground move in waves and go from left to right, and this time the sea drew back from the land until we were stranded and the bottom of the sea was exposed, so that we saw what had never been seen before, fish struggling on the sea-bed and the monsters of the deep aground. The round-hulled ships rolled over on their sides, while our *Wateree* sat down upon her flat bottom; and when the sea came back, returning not as a wave, but rather as a huge tide, it made our unhappy companions turn turtle, whereas the *Wateree* rose unhurt on the churning water.

From that moment on, the sea seemed to defy all natural laws. Currents rushed in opposite directions, dragging us along at a speed that we could never have reached even if we had been going at full steam. The earth was still quaking at irregular intervals, but less violently and for shorter periods each time.

The Peruvian ironclad *America,* which was held to be one of the fastest ships in the world at that time, was still afloat, and so was the American ship *Fredonia.* The *America,* who had tried to get out to sea with her engines running at full speed before the withdrawal of the water, was nevertheless partially stranded, and her hull was stove in. Now the sea was carrying her at a great speed towards the shore, and with her funnels belching thick clouds of smoke she seemed to be running in to the assistance of the helpless *Fredonia,* which was being drawn towards the cliffs of the Morro. Captain Dyer of the *Fredonia,* believing this to be the case, ran aft and hailed the man of war, which was now no more than a few yards away. "Ahoy! You can do nothing for us, our bottom is smashed in. Save yourselves! Good-bye!" A moment later the *Fredonia* broke to pieces against the cliff and not a man was saved, while a countercurrent miraculously took hold of the Peruvian ship and carried her in the other direction.

The last rays of the sun were lighting up the Andes when we saw to our horror that the tombs in which the former inhabitants had buried their dead, in the slope of the mountain, had opened, and in concentric ranks, as in an amphitheater, the mummies of natives dead and forgotten for centuries appeared on the surface. They had been buried sitting up, facing the sea. The nitre-impregnated soil had preserved them astonishingly, and the violent shocks that had crumbled the desert-dry earth now uncovered a horrifying city of the dead, buried long ago.

Words cannot convey the appalling appearance of the scene. Our minds had been much worked upon by what we had undergone already and we were ready to believe that the Day of Judgment had come and that the world was going to disappear: the bitterness of so terrifying a death went beyond anything that we could imagine.

It had been dark for some time when the lookout hailed the deck and said that a breaking wave was coming. Staring into the night we first made out a thin phosphorescent line which, like a strange kind of mirage, seemed to be rising higher and higher in the air: its crest, topped by the baleful light of that phosphorescent glitter, showed frightful masses of black water below. Heralded by the thunder of thousands of breakers all crashing together, the tidal wave that we had dreaded for hours was at last upon us.

Of all the horrors, this seemed the worst. We were chained to the bed of the sea, powerless to escape; we had taken all the precautions that were humanly possible, and now we could do nothing but watch this monstrous wave approach, without the moral support of having something to do or the hope that the ship could go through the mass of water rushing to overwhelm us. We could only hold on to the rails and wait for the catastrophe.

With a terrifying din, our ship was engulfed, buried under a half-liquid, half-solid mass of sand and water. We stayed under for a suffocating eternity; then, groaning in all her timbers, our solid old *Wateree* pushed her way to the surface, with her gasping crew still hanging on to the rails. A few men were seriously hurt; none was killed and nobody was missing. It was a miracle that I can scarcely really believe in even at this length of time.

Our survival was certainly due to the construction of the ship, her shape, and her fitting out, which allowed the water to pour off the deck almost as quickly as if she had been a raft.

The ship had been carried along at a very great speed, but all at once she became motionless. In the end, after a short wait, we lowered a lantern over the side and we discovered that we had run aground. Where we were we could not tell. There were still a few waves that came to strike us, but they were not so strong, and presently they stopped altogether. For some time we stayed at our posts, but as the ship remained quite still and nothing further happened the order was given for the exhausted crew to go below and sleep in their hammocks.

The sun rose upon such a spectacle of desolation as c
have been seen. We were high and dry, three miles from
chorage and *two miles inland.* The wave had carried us at
believable speed over the sand dunes which line the shore, a
valley, and beyond the railway line that goes to Bolivia, leavi
at the foot of the coastal range of the Cordillera of the Andes. Upon
an almost vertical cliff we found the mark that the tidal wave had
left: it was forty-seven feet up. If the wave had carried us on for
another sixty yards, it would have smashed us against the perpendicular mountain wall.

Near us there lay the wreck of a big English three-master, the
Channacelia; one of her anchor chains was wrapped round her as
many times as its length would allow, thus showing how the vessel
had rolled over and over, head over heels. Some way further off,
nearer the sea, lay the ironclad *America* upon her side, quite
wrecked.

During the days that followed the earth went on shaking, but none
of the tremors reached either the violence or the length of the first;
yet some were still quite strong enough to make the *Wateree* rattle
like an old kettle, and we had to leave the ship and go and camp on
the plateau, two hundred feet higher up. From that height we could
see the disastrous effects of the earthquake upon the topography. In
some places we found enormous fissures, some of which were more
than a hundred feet wide and of an unknown depth, while others
were no more than ordinary cracks. Both the one kind and the other
showed how panic-stricken the people must have been when they
fled: for example, I remember seeing the body of a dead woman on
the dead body of her horse, both having been swallowed by a crevasse
as they were flying for their lives.

The town itself had disappeared: where it had stood there stretched
an even plain of sand. Except in the suburbs on the mountain slopes
there was not a single house to show where Arica had been. All the
buildings, which were made of those hollow bricks called *adobes,*
had been destroyed by the shocks, and then the debris had been
swept away by the sea. In the suburbs that lay above the level that
the sea had reached we walked over a hideous piling up of everything,
including corpses, twenty or thirty feet deep.

Out of Arica's ten or fifteen thousand inhabitants, a bare few
hundred had survived. For the three long weeks during which we
waited for help, we shared the *Wateree*'s victuals and drinking water
with these wretched people. I will not attempt to describe our feelings
when at last we saw the old United States Navy frigate *Powhatan*
come into the roadstead, with her holds and her deck overloaded
with all possible kinds of victuals and stores.

Some few writers (the merest handful in the scientific fields), have stated that tsunami may not always be of low height on the open sea, an opinion periodically reinforced among certain segments of the public by the film industry's efforts to dramatize such waves. So preponderant is the opposite belief, notwithstanding, that one might tend to dismiss these claims out of hand were it not for the fact that they are occasionally made by authorities of distinguished reputation.

For example, in his book, *Chains of Fire—The Story of Volcanoes*, pages 114–115, Kent Wilcoxson writes of the third Krakatoa tsunami as follows: "Shortly after 10:02 A.M., and the greatest explosion, an immense wall of water towering *fifty feet high on the open seas* bore down on the coasts of Sumatra and Java . . ." The italics are mine, but the words are clear and unmistakable.

Again, in support of Wilcoxson, we have the following quote from *The Oceans—Their Physics, Chemistry, and General Biology* by Sverdrup, Johnson, and Fleming, page 543: "Destructive waves caused by earthquakes, dislocation waves, or tsunami are in general associated with submarine landslides which directly create transverse waves. These waves may reach enormous dimensions both *in the open sea* and near the coasts, and they proceed as ordinary long gravitational waves." Again, the italics are mine.

Regardless of the fact that one might dispute the above claim concerning "submarine landslides" (which appear to be *not* the causative factor in the generation of most tsunami, but responsible for only a minority of them), the text is in general a good one and its authors are worthy of respect. But where do they and the few who agree with them get the idea that seismic sea waves may reach "towering" heights or "enormous dimensions" on the *open sea*? To quote again from Wilcoxson: "The keeper of the lighthouse at Vlakke Hoek [about 40 miles west of Krakatoa] reported that the water rose fifty feet. Since this position was exposed on the point of a peninsula, it is taken that fifty feet was the height of the wave on the open sea."

Viewed logically, such a conclusion, based on presumption rather than any direct evidence, seems scarcely tenable—while the authors of the other text mentioned neglect, in this specific instance, to give their source of information. But then, how can one explain the report of the Dutch lighthouse keeper? Approaching the matter logically, it seems probable that when he saw the wave it *was* 50 feet high. The Sunda Strait is part of the South China Sea and is very shallow, as any chart of it showing soundings in fathoms will readily attest. Again, the distance from Krakatoa to Vlakke Hoek is a mere 40 miles. Isn't it obvious then that the wave, caught in the restricted confines of this narrow body of water, "feeling" the bottom starting to shoal before it was more than minutes old, would begin to lift while still a considerable distance from its ultimate point of impact? From the high vantage point of the lighthouse keeper, it must have appeared that the entire ocean had risen 50 feet and was moving against that part of the coast he occupied. In reality, far from being 50 feet high *on the open sea* in the sense used here, the tsunami, having encountered the shoreline's rising gradient at the time it was observed, was already slowing its front and increasing its height in the usual manner of its kind.

Incidentally, for the vulcanologist, Krakatoa today offers a very interesting study. As the pre-eruptive mountain of 1883 was one that had risen from the great caldera left by the cataclysmic explosion of May of 1680, so the presently existing cinder cone has risen from the ocean-filled crater of the more recent eruption. In the Indonesian tongue this relatively small prominence in the middle of its quiet lagoon is called Anak Krakatoa, meaning "Child of Krakatoa." It first appeared following mild eruptive activity in 1929, when what had been a blast-created hole nearly 1,000 feet under water completed the build-up process of converting itself into a lifeless island of black pumice and cinders. Today, it continues to rebuild. The last observed eruption was on January 12, 1960. And nothing has changed. The crust-splitting zones of shear continue to knife across the Sunda Strait. Anak Krakotoa smokes and occasionally rumbles from within its furnace. Very gradually the gases continue to form, the subterranean pressures to increase.

4

Alaska and the Hawaiian Islands

The Aleutian Trench: April 1, 1946.   Unimak Island and the Scotch
Cap Lighthouse.   Hawaii: Hilo and the Waipio Valley.   Kawela Bay:
Another eyewitness description.   Observations based on
fluid dynamics

The date was March 31, 1946: the time, 11:50 P.M. At the eastern
end of the Aleutian chain Unimak Island lay in black, dismal iso-
lation, its ironbound coast pounded by endless surges of the sea,
its desolate, partially ice-locked reaches dominated by the smoke-
plumed, towering immensity of Shishaldin Volcano. At the southwestern tip of the island the Scotch Cap Lighthouse stood stark
against the night sky, located there to warn ships of the dangerous
approaches to Unimak Pass. The original 45-foot structure, built
in 1903, had boasted a wooden octagon for a tower. In 1940 it was
torn down, its place taken by a square, all-white building of steel-
reinforced concrete, a square tower jutting from its center, a huge,
metal-latticed beacon mounted atop the tower and rising to 60 feet.
The foundation of the building was at an elevation of 46 feet above
mean low water, its 80,000-candlepower light 98.4 feet above the
same mark, the higher of its two radio masts 108 feet above it. All
in all, it was an imposing edifice, far more solid and durable than
the cluster of momentarily unoccupied dwellings that flanked it on
the eastern side, a monument to modern lighthouse engineering
well suited to serve the volcano-studded island which Russians of
the preceding century called "The Roof of Hell."

Five men manned the light that night, all of them veterans of the U.S. Coast Guard. They were Chief Boatswain's Mate Anthony Pettit from Hancock, Michigan; Fireman First Class Jack Colvin from Juneau, Alaska; Seaman First Class Dewey Dykstra from Artesia, California; Seaman First Class Paul Ness from Seattle, Washington; and Motor Machinist's Mate Second Class Leonard Pickering from Shreveport, Louisiana.

At 11:50 P.M. a card game was nearing its end. At the top of the cliff and slightly inland was a D/F (direction-finding) station, and one of its personnel had joined the Scotch Cap off-duty light crew to while away the time. Just at midnight he played his last hand with them, deciding to turn in. They offered him a bunk for the night. He declined with thanks, then made the trip back to his more lofty quarters. Just then, at both the lighthouse and the D/F station, the men of the midwatches relieved their mates and came on duty. It was now April 1.

For an hour and a half nothing occurred to disturb the monotonous routine. Another card game started. The sounds of the night were the usual ones, the all-pervading crash and recession of the sea as surf beat endlessly along the cliffs. Then, at precisely 0130 hours, the hands at both stations experienced a violent earthquake lasting more than half a minute, a tremor of unknown magnitude. On the clifftop the D/F operator made the following entry in the log:

0130—Severe earthquake felt. Building rocked severely. Objects shaken from locker shelves. Duration approximately 30–40 seconds. Building creaked and groaned but no apparent damage. Weather clear, calm.

At both stations the men were tense now, talk continuing in nervous tones as lighthouse and D/F station compared notes by telephone, discussing similar experiences. They tried to reassure one another. All along the Aleutian archipelago low-magnitude shocks were frequently recorded, nearly as often as in Japan. This was scarcely different. Oh, a little bigger, perhaps, but nothing to cause alarm. Or was it? In the private world of his own inner thoughts, each of the men knew better. It was much bigger. Among some of the hands the usual wisecracking increased but it seemed to have a false ring to it.

In reality, their growing apprehension was more than justified. While they chatted, distant seismograph stations were to locate the epicenter on the north slope of the Aleutian Trench at 53°N, 163°W, a position slightly over 90 miles to the southeast of Scotch Cap. The ocean here is 2,300 fathoms deep, but the bottom shoals very rapidly to the north, and in the latitude of the Aleutian Islands the water already acquires the characteristic shallowness of the Bering Sea, with an average depth of less than forty fathoms. In this area of extreme seismic instability a high-magnitude earthquake had struck, weakening one wall of the mighty trench which knifed so deeply into the earth's crust, fracturing it in depth. The stage was set for tragedy. Crushing disaster was imminent.

Twenty-seven minutes went by. Then a second great shock passed at high speed beneath the ocean floor, smashing at the Aleutians all along their length, causing the ground at Unimak to undulate savagely. The D/F operator made another entry:

0157—Second severe quake felt. Shorter in duration but harder than at 0130. . . . Again, no apparent damage, although building shook severely.

Scotch Cap phoned the D/F. Once more reactions were compared, words of encouragement exchanged. No, it would take more than that to harm the lighthouse. It was really built, you know, really built. That last one was big, though, no question about it.

It was big. Some 13,800 feet down, the wall of the Aleutian Trench bulged ominously as the ponderous fissure widened. Then an enormous piece of the trench wall ripped loose and millions of tons of rock and earth plunged dizzily downward. Titanic shock waves were generated in the sea, their colossal energy racing toward the surface at fantastic speed, there to create tsunami which now moved outward in ever-widening concentric circles that would ultimately span the Pacific. Unimak Island was in the path of advance. Closest to the epicenter of the disturbance, it would be struck first.

On the open sea the great wave was perhaps 2 feet in height. However, that segment of its northern arc which rolled toward Unimak was moving at some 5 miles a minute, or 300 miles per hour. The distance it had to traverse was short, a mere 90 miles.

Encountering rapidly shoaling water it would begin to slow down, its speed dropping to 200, then to 150 mph, until, by the time it reached the Davidson Bank with its average depth in this area of 41 fathoms (246 feet) its velocity would decrease to 72 mph.[1] It would strike the island at about that speed. Taking note of the depth alterations over the track it traversed, calculating average depth, its arrival could be expected in 21 minutes.

The men on Unimak knew nothing of this. None had ever experienced a seismic sea wave. If the vague memory of having heard of such things passed through the mind of one of them and was mentioned, the fact was not recorded. They didn't know what to expect. Significantly, the joking had stopped as nerves began to wear thin. Conversation lagged. No one seemed to recall that this was April Fools' Day.

Speeding across Davidson Bank in the infernal blackness which concealed its passage, the wave was no longer what it had been. Now in shoal water, it responded, according to a natural law of physics, to the friction exerted on it by the rising sea bottom: its front slowed down, while the waters still racing behind it, blocked by the decelerating front, lifted higher and higher. It was at least 100 feet high now, a watery colossus blacker than the night itself, silvery white along its vast crest, elemental, irresistible, terrible. As an avalanche displaces air in its mighty fall, altering atmospheric pressure sufficiently to collapse the walls of houses beyond the reach of its snow and rock, so, on a far greater scale, did the wave. Under its lofty curl, air pressure was continuously changing as rapidly moving air molecules sought escape from the virtually incompressible mass of liquid, trillions of them rushing this way and that in violent collision, augmenting the sound of molecular displacement within the water itself. It was the rising roar thus created which became audible to the men on the island. At 0216 they heard it even through the heavy plate glass of the lighthouse. Above, on the cliff, D/F personnel heard it too. Faces were turned toward the sea, eyes straining to pierce the dark. All unknowing, helplessly they waited.

High up, the men of the D/F station heard the incredible din

---

[1]These figures are derived by substituting into the formula for tsunami velocity. See Chapter 3, page 35.

rise suddenly to earsplitting levels as with a stupendous, climactic detonation the tsunami burst against the island's black ramparts of rock. Like a deafening sonic boom was the way the men described it later. The headland on which the lighthouse was built was totally engulfed. The incalculable power of millions of tons of water struck with a force far exceeding that of nitroglycerin, quite literally pulverizing all the matter subject to the impact, sweeping upward and beyond in a lightning surge that burst over the cliff's edge and swirled in a wild torrent about the D/F station.

The latter's personnel were stunned. Crowding into the operations office, they asked their commanding officer for orders and were told to get at once to higher ground. As they started to climb, someone happened to look down. Then it was discovered. In the direction of Scotch Cap no light was to be seen. The huge beacon had been extinguished. The foghorn was silent. It was quickly determined that all radio and telephone contact was broken—dead. In subdued tones, some of the men spoke of the frightful sound they had heard, then its culmination, like some earthshaking, thunderous explosion.

Remaining alone at the D/F station, the commanding officer wrote in the logbook:

> 0218—Heard terrific roaring of the sea followed by huge sea immediately. Top of wave rose above the cliff back of Scotch Cap light station and struck D/F station causing considerable damage. Crew ordered to high ground. Can't make radio contact so broadcast following message: PP NMJ NNF NND TIDAL WAVE X MAY HAVE TO ABANDON THIS PLACE X BELIEVE NNHK LOST INT R INT R KKK

NNHK was the code designation for the Scotch Cap Lighthouse, its presumed loss a grim report to have to send out. At this point the conclusion was hardly speculative. Although relatively certain of what had happened at the light, the CO had no way of confirming it until the dawn brought an end to this nightmare. *Five men,* he kept thinking. Five coastguardsmen down there below him where once there had been a light. As the night wore on, he could hear the heavy crash of other combers, the bigger ones arriving at precise intervals readily identified as part of the tsunami wave-train. Even these were now all smaller, of course. In this case, the first one had been the killer. But who could swear it was the last of those death-dealing titans? Who could say what might yet come out of the ocean's blackness? At intervals, the CO used a flashlight to make his way precariously to the cliff's edge, there to shine its beam downward. They were useless probings. The beam did not reach far enough. He saw nothing, heard no sound but the endless send and surge of the sea. Feeling hopelessness grip him, he wrote again in the log:

0245—Scotch Cap believed lost; lights extinguished and horn silent.

Shortly before 3:00 A.M. the phone rang. It was the crew at Cape Sarichef calling, the island's other Coast Guard lighthouse at the north end of Unimak Pass. The CO told them what had happened and what he suspected, giving them pause to reflect on how much safer they were on the Bering Sea than on the Pacific. He hung up the phone, then plunged his men into activity. A few minutes later he turned to the log once more and wrote:

0300—Gathering emergency stores and clothing. Called crew back (as sea is apparently abating) to get more clothing and canned goods. Emergency lights rigged and carrying clothing and supplies pack.

The D/F personnel were together again, all of them shivering with cold, trying desperately to warm themselves. Slowly the night passed, with infinite slowness, it seemed to those who waited. When at last the dawn came to gradually disperse the darkness, they went to the rim of the precipice and looked down. What they saw nearly froze the blood. The Scotch Cap Lighthouse was gone. Where it had stood was an exposed cement foundation littered with shattered concrete and weirdly twisted steel; girders, ties, and trusses, all of the same stout metal, either bent flat or wrenched out of their sockets and missing, obliterated.

They climbed down to search the slope for bodies. Debris was everywhere, scattered for hundreds of yards above the building site as though the structure itself had exploded. Right to the level of the plateau (118 feet above mean low water), the slopes had been ground and scoured by the wave's frightful impact, gouged so deeply as to expose the stratification of the underlying rock structure.[2] Of

---

[2]This is clearly seen in Coast Guard photographs.

the five doomed men only a small and pitiful remnant was found. The log made poignant note of the fact:

0700—Went to light station; debris strewn all over place. Piece of human intestine found on hill.[3]

That was what a tsunami could do. The CO knew it was time to send a teletype report of the disaster to all Coast Guard ships and stations. He dictated the following:

TIDAL WAVE PRECEDED BY EARTHQUAKE COMPLETELY DESTROYED SCOTCH CAP LIGHT STATION WITH LOSS OF ALL HANDS X TOP OF WAVE STRUCK THIS UNIT CAUSING EXTENSIVE DAMAGE BUT NO LOSS OF LIFE X ENGINE ROOM FLOODED OUT BUT EXPECT TO BE ABLE TO MAINTAIN EMERGENCY POWER IF NO FURTHER DAMAGE EXPERIENCED X WILL SEARCH FOR BODIES OF SCOTCH CAP PERSONNEL AS SOON AS POSSIBLE X REQUEST INSTRUCTIONS.

The search was made, the members of the direction-finding station carrying out a joint effort to find some sign of the missing coastguardsmen. They found nothing, not even shreds of clothing or personal effects. A subsequent search days later produced melancholy results when what was left of two of the men was at last discovered, broken remains far from the building site. Of the other three no trace was ever found.

Today, a tablet of bronze commemorates the loss of the five servicemen. There is again a lighthouse at Scotch Cap. A makeshift and temporary installation was succeeded in 1950 by a permanent structure built 500 feet above the original foundation. This was a manned facility, but in 1971 the crew was transferred elsewhere when the light was redesigned to make it automatic.

How easily people forget. Though lightkeepers at those stations which are still manned will remember Scotch Cap for years to come, that tiny section of the general public that was ever aware of it has long since consigned its memory to limbo. What was it like that night? What was it *really* like? Perhaps artist Darrell Millsap came

---

[3]Log entries, teletype reports, and other pertinent data courtesy of U.S. Coast Guard Archives Section, Coast Guard Headquarters, Washington, D.C., and U.S. Coast Guard Academy Library, New London, Connecticut.

close to reality when he painted the awesome event from the Coast Guard report augmented by imagination. No one can look at his dramatic oil and be indifferent to it. The boiling sea is in the foreground, white as it foams round the rocks. The lighthouse is in the middle distance. In the background is the wave, faintly illuminated, a pitiless monster of nature towering over the scene like something out of the bowels of hell. The big beacon light plays on the tsunami, its yellow beam holding steadily on the curling center. On the ground floor, other lights shine from windows.

But the viewer senses the hopelessness, knowing positively that nothing on earth can save the occupants of the doomed building. Nothing.

Four hours and 34 minutes after the second temblor ripped the Aleutian Trench that April 1, the first tsunami reached Oahu in the Hawaiian Islands, having averaged 490 mph over a distance of 2,240 miles. Considering its route across the Pacific, it might be expected to have struck Windward Oahu (the island's east coast) with as much fury as the North Shore, but this would be to overlook the all-important factor of shoreline configuration. Actually, so far as Oahu is concerned, the waves were most destructive at the north shore's Kawela Bay, where an excellent photograph of the "nonbreaking" type of tsunami was taken by Dr. Francis P. Shepard, a marine geologist employed by Scripps Institute. In his book, *The Earth Beneath the Sea,* the scientist writes of his experience in one of this century's few eyewitness accounts:

> . . . we were sleeping peacefully when we were awakened by a loud hissing sound, which sounded for all the world as if dozens of locomotives were blowing off steam directly outside our house. Puzzled, we jumped up and rushed to the front window. Where there had been a beach previously, we saw nothing but boiling water, which was sweeping over the ten-foot top of the beach ridge and coming directly at the house. I rushed and grabbed my camera, forgetting such incidentals as clothes, glasses, watch, and pocketbook. As I opened the door I noticed with some regret that the water was not advancing any farther but, instead, was retreating rapidly down the slope.
>
> By that time I was conscious of the fact that we might be expe-

riencing a tsunami. My suspicions became confirmed as the water moved swiftly seaward, and the sea level dropped a score of feet, leaving the coral reefs in front of the house exposed to view. Fish were flapping and jumping up and down where they had been stranded by the retreating waves. Quickly taking a couple of photographs, in my confusion I accidentally made a double exposure of the bare reef. Trying to show my erudition, I said to my wife, "There will be another wave, but it won't be as exciting as the one that awakened us. Too bad I couldn't get a photograph of the first one."

Was I mistaken? In a few minutes as I stood at the edge of the beach ridge in front of the house, I could see the water beginning to rise and swell up around the outer edges of the exposed reef; it built higher and higher and then came racing forward with amazing velocity. "Now," I said, "here is a good chance for a picture." I took one, but my hand was rather unsteady that time. As the water continued to advance I shot another one, fortunately a little better. . . . As it piled up in front of me, I began to wonder whether this wave was really going to be smaller than the preceding one. I called to my wife to run to the back of the house for protection, but she had already started, and I followed her just in time. As I looked back I saw the water surging over the spot where I had been standing a moment before. Suddenly we heard the terrible smashing of glass at the front of the house. The refrigerator passed us on the left side moving upright out into the cane field. On the right came a wall of water sweeping toward us down the road that was our escape route from the area. We were also startled to see that there was nothing but kindling wood left of what had been the nearby house to the east. Finally, the water stopped coming on and we were left on a small island, protected by the undamaged portion of the house, which, thanks to its good construction and to the protecting ironwood trees, still withstood the blows. The water had rushed on into the cane field and spent its fury.

My confidence about the waves getting smaller was rapidly vanishing. Having noted that there was a fair interval before the second invasion (actually fifteen minutes, as we found out later), we started running along the emerging beach ridge in the only direction in which we could get to the slightly elevated main road. As we ran, we found some very wet and frightened Hawaiian women standing wringing their hands and wondering what to do. With difficulty we persuaded them to come with us along the ridge to a place where there was a break in the cane field. As we hurried through this break, another huge wave came rolling in over the reef and broke with shuddering force against the small escarpment at the top of the beach. Then, rising as a monstrous wall of water, it swept on after us, flattening the cane field with a terrifying sound. We reached the comparative safety of the elevated road just ahead of the wave.

. . . we walked along the road until we could see nearby Kawela Bay, and from there we watched several more waves roar onto the shore. They came with a steep front like the tidal bore that I had seen move up the Bay of Fundy at Moncton, New Brunswick, and up the channels on the tide flat at Mont-Saint-Michel in Normandy. We could see various ruined houses, some of them completely demolished. . . .

Finally, after about six waves had moved in, each one apparently getting progressively weaker, I decided I had better go back and see what I could rescue from what was left of the house where we had been living. After all, we were in scanty attire and required clothes. I had just reached the door when I became conscious that a very powerful mass of water was bearing down on the place. This time there was simply no island in back of the house during the height of the wave. I rushed to a nearby tree and climbed it as fast as possible and then hung on for dear life as I swayed back and forth under the impact of the wave. Like the others, this wave soon subsided, and the series of waves that followed were all minor in comparison.

"Well," I thought, "you're a pretty poor oceanographer not to know that tsunamis increase in size with each new wave." As soon as possible I began to look over the literature, and I felt a little better when I could not find any information to the effect that successive waves increase in size, and yet what could be a more important point to remember? You can be sure that since then those of us who have investigated these waves in the Hawaiian Islands have stressed this danger. . . .

The waves struck other islands of the Hawaiian group, falling with particular force on the northeast shore of the "Big Island" itself, the Island of Hawaii. Here, about 17 miles from the westernmost outthrust of Upolu Point, lies the great scenic gash in the pali coast called Waipio Valley. The gap is 6 miles long, winding downward to reach a maximum gradient of 26 percent. It may certainly be described as beautiful. The lush verdure is threaded by a stream, one of several draining the mile-high Kohala Mountains. Guavas are plentiful, and *noni* apples, taro and ginger and many species of fern. And history was made here. In 1780, a gathering of chiefs recognized Kamehameha as future king. But the waves of 1946 were not to spare this lovely place.

In all likelihood, the same tsunami which destroyed the Scotch

Cap Lighthouse (but a different portion of its ocean-girdling arc) was the one that inundated the valley. It was believed to have reached a height of over 50 feet in this area. As it thundered inland, the homes of the settlers were smashed to bits, their taro farms swept away, tens of thousands of tons of rich topsoil removed as by the scouring action of a gigantic knife. Today, there are few traces of the disaster. Nature has a way of quickly healing such scars.

Another point which felt the blow was the city of Hilo, on the same coast, slightly more than 40 miles to the southeast as the crow flies. The highest crest on Oahu was recorded at 35 feet, but here on Hawaii it crested in at least one area (Polulu Valley, ten miles northwest of Waipio) at 54 feet. As mentioned in the preceding chapter, Hilo was hit hard. From the deck of a ship in Hilo Bay the U.S. Army Corps of Engineers took a photo. It shows the tsunami crushing a warehouse, one of many destroyed in the dock area. About to be engulfed is a solitary human figure, a man apparently taken by surprise, who was destined to die within seconds of the shutter's click.

To the islanders, seismic sea waves are not a phenomenon restricted to the twentieth century. From their discovery by England's Captain James Cook in 1778 until the year 1819, no record of them was kept in the Hawaiian Islands.[4] From that date until now, however, the islands have been hit no fewer than 41 times.

By now the reader may have noted that not all tsunami arrive at coastlines as great breaking waves. On the contrary, in many places they can take the form of a sudden and very rapid rising of the tide, with the inrushing water cresting far above the normal high tide mark, presenting a very turbulent front which nonetheless fails to break at the peak of its force but instead dissolves into foam at some distance inland. This was the case with the waves described by Dr. Shepard. It is also what happens in the majority of instances where a ship is picked up and carried inland and the damage to the vessel is slight. The final mighty wave at Arica seems to have been an exception. According to Lieutenant Billings's account, the lookout referred to this as a "breaking wave," and the officer's own

---

[4]In 1779, the great navigator was killed in a skirmish with natives at Hawaii's Kealakekua Bay, where his monument now stands.

description of it is in clear corroboration. How then did the *Wateree* survive it? We can only assume that the wave caught her just right; that, instead of suffering instant capsizement like the other ships in the bay, she lifted to the wave's thrust and was borne by it. This would also seem to be Lieutenant Billings's opinion.

This factor of significant *variance of form* between one seismic wave and another is characteristic. Since in matters of fluid dynamics tsunami obey the same laws as do wind waves, the difference between the two types is readily explained. Whether or not a wave breaks early or rolls on and on to dissipate its strength gradually depends partly on the form it takes at its birth, and, more particularly, on the topography of the shore it ultimately strikes. In the case of *wind waves* on the one hand, or true *tidal waves* on the other (the wavelengths of tsunami are intermediate between the extreme shortness of the former and the great length of the latter), experts in the field classify them as *spillers* or *plungers*. A spilling wave forms when its movement is over a gently sloping sea bed, gradually steepening until critical height is reached. At this point it begins to spill, after which portions or all of its front may be laced with continual cascades of foam. Such a wave may roll a long distance before it breaks, being the type favored by surfers and such outrigger enthusiasts as those who ply the waters near Diamond Head.

The plunger (called by Australian surfmen the "dumper") is formed when the wave encounteers a steep angle of slope. As it lifts abruptly to conform to the rapidly shoaling bottom, the onrushing water behind it piles up, pushing the crest higher and higher until the rising mass plunges forward to break with great violence.

Both of the above-described forms are taken by tsunami. However, where the spilling wind wave displays a smooth front, that of the late-breaking type of tsunami (as seen in the Shepard photograph) is usually very turbulent. (An exception will be cited below.) This type of wave, though the steepness of its forward face increases as it advances, may never "break" in the usual sense, but simply engulf the land in an irresistible, surging mass. Though the plunging type of tsumani may be the more spectacular of the two, being the traditional "tidal wave" of film and story, both can be deadly, and both are thoroughly worthy of respect. Needless to say, on arrival at coasts of varying topographical contour, the same wave can take *both* forms, as witness the Aleutians–Hawaiian Islands example in 1946.

It is, of course, obvious that no one directly in the path of the plunging type of tsunami ever survives its titanic impact. To the singular nature of the second type, however (which may be either a spilling wave or one that never crests at all), is due the centuries-long accumulation of some very colorful tales. Take, for example, the spine-chilling experience of Shigeto Matsubara, an employee of the electric company in the city of Hilo. On May 22, 1960, he was working at a power station down at the docks, and, though warned of the tsunami's approach, delayed his departure a little too long. As he was leaving, the wave caught him. In relating the strange occurrence to the National Geographic Society's William Graves, he reminisced as follows: "It didn't crest or roar like other big waves. I can only describe it as a sudden *presence* of water. And there we were, he wave and I, riding up one of the main streets with buildings going down on either side and with me floating upright high off the ground, like a maharaja on his elephant."

As similar waves had done with ships and boats in the past, so it happened with this incredibly fortunate Japanese gentleman. Having run its course, the tsunami set him down gently, from which point of "disembarcation" he rapidly fled. In the subsequent interview, Mr. Graves asked him if he'd been injured.

"Not by the wave," was the reply. "But I must have given such a terrible shout when it picked me up that I paralyzed my vocal cords, because for three months afterward I didn't have any voice."[5]

Concerning this 1960 tsunami at Hilo, a few words were written in Chapter 3. To them may be added the following, the result of a meeting between William Graves and Captain Robert C. Munson, at that time director of the U.S. Coast and Geodetic Survey's Honolulu field office in Hawaii and of its Seismic Sea-Wave Warning System. Graves recalled how he had visited Munson at the control center at Ewa Beach near Honolulu. There, with the aid of a large wall map of the Pacific, Munson had offered some interesting observations.

[5]Quoted from William Graves, from *Hawaii* (Washington, D.C.: National Geographic Society, 1970).

Seismic sea waves striking the Hawaiian Islands, he said, may be expected to come from the direction of Peru or Chile to the southeast, from Alaska to the north, or from Japan to the northwest. When they come from Japan they encounter what he referred to as "one of the world's longest and finest natural breakwaters," meaning Kure, Midway (of World War II fame), Lisianski, Laysan, and the other prominences of the Hawaiian archipelago, northwest of the main islands. Deflected by this volcanic spine, the waves coming from this direction have never yet proved a threat to the more heavily populated centers of the fiftieth state. To the north, east, and southeast, however, the approaches are all wide open.

Getting down to specifics in the case of the Island of Hawaii, Munson then added that Hilo is a natural target for tsunami, that because of its funnellike configuration the topography of Hilo Bay intensifies the destructive power of such waves. Engineers have, of course, advanced numerous ideas for the structural protection of the city, everything from the construction of a gigantic breakwater to the elevating of waterfront buildings on tall, rock-embedded rods of heavy steel. To date, all such plans have remained hypothetical, largely because of the prohibitive cost, a figure running into hundreds of millions. It's perhaps as well. In the event that such construction were to neutralize the effectiveness of some future wave of moderate size, it is logical to assume that thereafter the citizens would place full faith in it. But tsunami vary greatly in power and destructiveness. Whereas man's ingenuity might thwart one wave, or even several of them, Hilo's vulnerable location would remain the same, an immutable geographical fact. Eventually, a day would come, and *one particular wave*. A population relying on man-made protection, remaining in the city because of this reliance, would then perish.

In a general sort of way, a good deal has been said about the sheer brute power of seismic sea waves. However, after intensive research the author has been unable to discover any precise figures dealing with the actual mathematics involved in this power. This is not surprising. With other forms of wave, even the great storm waves that sink ships and ravage coasts, man has had virtually unlimited opportunities for study. Not so with tsunami. For one thing, their comparative rarity militates against any rapidly acquired familiarity. For another, it was not until the late 1940s that science had begun to take their study seriously.

Perhaps some idea of tsunami strength might be gleaned from a brief examination of the known power of storm waves in various areas, since this has been carefuly documented. Although numerous examples might be cited of the power of waves to hurl huge rocks high into the air, or to move them deep below the surface (orbital motion of internal waves), it is the proven effect of storm-wave action at sea level that is most impressive. On both sides of the Atlantic, for example, such wind-caused waves have destroyed famous lighthouses: the well-known landmark of England's Eddystone Light 14 miles south-southwest of Plymouth, for one; that of Minots Ledge (or Cohasset Rocks) fifteen miles southeast of Boston, for another. Both structures have been rebuilt not once but several times, despite repeated loss of life.

When we speak of wave action we're directly concerned with *foot-pounds of energy,* this being a unit of measurement which Webster defines as "being equal to the work done in raising one pound avoirdupois (as distinguished from troy weight) against the force of gravity the height of one foot." Consider, then, the following actual occurrences.

In the years between 1850 and 1855 a massive breakwater was built at Wick Bay in Scotland's northeast corner, an area of coast which receives the full force of the North Sea's November gales. For a foundation, into an enormous trench 100-ton granite blocks were laid, after which three tiers of ponderous boulders were cemented in, followed by an 800-ton upper layer of concrete waste material. Bound together by hundreds of iron ties measuring nearly four inches in diameter, the entire mass represented a feat to give pride to engineers and was a monument to the ingenuity of man. Then came the mighty storm of 1872. Before it was over, 195 feet of breakwater was carried away or demolished, including 1,350 tons of upper-layer cement and lower-tier blocks. The entire outer end of the structure, 11 feet high and 45 feet in length, simply split apart and went under.

The builders tried again. With great effort they repaired the damage, brought other boulders into place, tied them down with hundreds of tons of cement and more iron, and replaced the destroyed 800-

ton cap with one of 2,000 tons. All to no avail. In 1877 another great storm struck, again reducing the breakwater to a total wreck and removing large parts of it outright. What was the estimated energy required to accomplish this? R. C. H. Russell, the noted British authority on waves, calculated that it could not have been less than 2,440 pounds pressure per square foot. According to his estimates, the 40-foot storm waves often seen at Wick could easily develop pressures exceeding 4,000 pounds per square foot.

If such pressures seem startling, they may appear less so in the light of comparative phenomena. For example, take the case where a storm wave breaking against a cliff or man-made seawall does so in such a way as to trap an air pocket and compress it. If the pocket is relatively symmetrical (i.e., nearly as thick in one dimension as the other), it will absorb the impact. But if it is thin (i.e., if its height is much more than double its thickness), shock waves will be generated within the air mass, being greatest where the pocket is thinnest. In such cases, the compression of the air molecules instantaneously exceeding the critical point, the air will actually *explode,* hurling booming sheets of water straight up in high speed jets of spray. Again, according to Russell, the pressure developed by such blasts of violently displaced fluid has been estimated at 16,800 pounds per square foot, far more than enough to instantly decapitate a man whose head happened to be in the way.

If such pressures as these may be generated by 40-foot storm waves, what might be the energy coefficient in a 100-foot tsunami? What must it have been in the Cape Lopatka (Kamchatka Peninsula) wave of 1737, a phenomemon 210 feet in height? The mathematical probabilities stagger the imagination. Yet tsunami much larger even than this have occurred (see Chapter 10) and will occur again.

The reader will recall Lieutenant Billings's description of the culminating wave at Arica, the great breaking tsunami that struck after the onset of darkness. "Staring into the night we first made out *a thin phosphorescent line* which, like a strange kind of mirage, seemed to be rising higher and higher in the air: its crest, topped by *the baleful light of that phosphorescent glitter,* showed frightful masses of black water below." The author has provided the italics, wishing to focus attention on one of the most curious and fascinating manifestations associated with seismic sea waves, that of their occasional luminosity when seen at night.

How often down the centuries has the terror inspired by such waves been augmented by superstitious dread? Throughout many areas of the Pacific Basin the legends of primitive tribal societies speak of "the waves which shine in the dark," and some hint of this singular characteristic may be observed along the crest of the Unimak tsunami in Darrell Millsap's painting.

What occurs in nature to produce such effects? Actually, as most seismologists would attest, displays of luminescence have often been seen both before and during large earthquakes, some of them attributable to lightning storms, but others less easily explained. In the case of the tectonic shocks of great magnitude the variety of such phenomena is often amazing, ranging from glowing clouds (not to be confused with the deadly *nuée ardente* of the Peléan type of volcanic eruption) and sudden fiery flashes to exploding fireballs and dazzling rods or pillars of shimmering white light. To many of the questions posed by these occurrences science does not as yet have the answers. But one thing seems clear. Though earthquakes of a particular type (and the resultant crustal displacement) are the chief cause of tsunami, there is no causal relationship between wave luminescence and the weird atmospheric effects cited above.

To what then is wave luminescence attributable? Bearing in mind that such effects are seen only at night, and then only with a *minority* of tsunami, it may be said that the bulk of the evidence points to either (or both) of two possible causes. For a long time the first cause, that of quantities of phosphorous in the water, was believed by many oceanographers to be the only one. However, thanks to the protracted and intensive studies undertaken by marine biologists and others, especially Japanese, a second possible cause has been isolated, this one biological as well as chemical. For centuries the Japanese were aware that night tsunami striking the east coasts of the home islands occasionally gave off a strange light. Some studies were begun. Theories were propounded. Yet no definite answers were forthcoming. Then came the year 1933 and another of the great undersea earthquakes to which the Sanriku Coast (northeast Honshu) has often fallen victim. The temblor struck at night. Within minutes, through the darkness covering the sea, residents fleeing the waterfront area witnessed the most terrifying display of oceanic pyrotechnics any of them could have imagined. The incredibly vivid flashes of light were emitted from incoming tsunami, an attendant feature with each of the waves which rolled that night against Kamaishi Bay.

Japanese scientists renewed their quest, redoubled their efforts. Ultimately, their biologists began to study an all-but-forgotten marine creature, a very minute organism called *Noctiluca miliaris*. This animal swarms in the sea by the uncounted trillions. Of the genus *Noctiluca* (Latin for "something that shines by night") and the order *Dinoflagellata,* in scientific parlance it is designated as belonging to a genus of marine bioluminescent flagellates, meaning simply that it gives off light (due to the presence of organic phosphorous) and is equipped with whiplike appendages (flagella) for swimming.

Is this the chief cause of tsunami luminescence? Almost certainly it is. For one thing, free phosphorous in the sea would scarcely be found in sufficient concentration to account for the intensity of the radiance observed. For another, *Noctiluca miliaris* (along with certain other organisms of the same genus) has definitely been established as a primary contributing factor in marine phosphorescence. Why their presence in certain tsunami? Apparently, disturbed by the earthquake shock, they begin to swarm, and are then picked up in their myriad numbers to be concentrated in the wave's advancing crest. Sometimes they are present in the area of disturbance, sometimes not. When they are, they become the stuff of legends.

5

The New Global Tectonics

Crustal plate slippage and disaster in South America. Chile:
May 21, 1960.

While few scientists today accept Alfred Wegener's theory of continental drift in its original form, a significant majority subscribe
to it in one or another of its subsequent modifications. One of those
who arrived at this position in 1960 was the eminent Canadian
geophysicist, J. Tuzo Wilson, formerly Director General of the On-
tario Science Centre and Past President of the International Union
of Geodesy and Geophysics. Writing for the *National Geographic*,
Dr. Wilson pointed out that not until after 1965 had sufficient
evidence been amassed to give firm support to the long neglected
and much criticized Wegener hypothesis. This changed attitude
on the part of the international scientific community stemmed primarily from the post–World War II study of paleomagnetism (magnetic records in ancient rock detected in this instance by ships'
magnetometers), and the ultimate plotting of the bi-hemispheric,
40,000-mile-long Mid-Oceanic Ridge, both of which provided science with positive indications that the continents are actually in
motion.

Today, the theory of a hard and brittle crust floating on a highly
viscous mantle is often called that of plate (or global) tectonics.
This envisages a mantle only the upper 300 miles of which is
sufficiently plastic to permit flowage, however slowly, of a layer
called the asthenosphere, from the Greek *asthenes*, meaning "weak."
Everything above that—the very top of the mantle and the crust·

in its entirety (comprising together a total thickness of about 100 miles) is termed lithosphere, from *lithos*, meaning "stone."

Due to subterranean heat, the pressure of expanding gases, and fluid movement within the mantle (theory of convection currents), the lithosphere is everywhere fractured both latitudinally and longitudinally, the resultant crustal divisions each bearing its distinctive geographical name and differing markedly from the others both in size and shape. These divisions are called *plates*. At present, there are six of major size throughout the world, with at least twelve of lesser scope thus far identified. The major plates bear the following names: Pacific, Indo-Australian, Antarctic, American, African, and Eurasian. Among the minor plates are the Philippine, Cocos, Nazca, Caribbean, Somali, and Arabian.

Though geologists see plate movement as indescribably gradual, they nonetheless view it as sure. Thus, these enormous sections into which the earth's thin crust is divided shift according to the sympathetic impulses of the asthenosphere's convection cells flowing beneath them. In areas where their broken edges collide, earthquakes are the usual result. Where they fracture, they permit the effusion of sub-crustal molten material, creating ridges (such as the East Pacific Rise), mountain ranges, and volcanoes. When one plate overwhelms another (the geologic term is *subduction*), the lesser of the two is driven down into the mantle (usually at one of the pre-existing zones of weakness called a *trench*), there to be again melted and absorbed.

Is it possible to detect the meeting of two such plates with the naked eye, to actually distinguish topographical features giving visual evidence of their edges being in contact? The answer is yes. To select an example with which many Americans are familiar, we may take California's famed San Andreas Fault, the fissure responsible for the disastrous San Francisco Earthquake of 1906. From the air, this great crack may be easily observed as it runs through desert country north of Los Angeles. Because of clarity of definition, it is often photographed in the extremely arid wasteland of the Elkhorn Scarp, where a view looking north puts the Carrizo Plain on the fault's left and the Elkhorn Plains on its right.

Let's consider this a moment. The San Andreas Fault is approximately 700 miles in length, running all the way from Cape Mendocino to just below the Mexican border. When we look at an aerial photograph of it, we are actually seeing the line that marks *a segment* of the conjunction of two of the earth's major tectonic plates, the Pacific to the west and the American to the east, though the grinding action of the ages has filled the fault with rock debris and makes it appear shallow. Science recognizes four types of plate movement, the first three being characterized by sea-floor spreading (crustal expansion), continental collision (crustal upthrusting), and subduction (crustal override). The fourth is lateral transformation, where one plate slides past another moving in the opposite direction. In Chapter 2, this phenomenon was mentioned in relation to the Mendocino Fracture Zone, though millions of years may pass before this enormous fissure becomes extensive enough to subdivide the Pacific Plate. Into this category may be placed the San Andreas Fault, since the Pacific Plate is gradually moving to the northwest and the American to the southeast. Thus, the San Andreas system is called a transform fault.

If from the coast of southern California we move southward into the Pacific, we come to the Cocos Plate, the crustal division whose northeastern edge flanks most of Mexico and Central America, and demarcates a subduction zone with the Caribbean Plate's western extremity. South of this is another, much larger plate, the Nazca, extending westward into the Pacific to approximately the longitude of San Diego and embracing nearly the full length of South America. This Nazca Plate is geologically fascinating. Together with that part of the American Plate that includes the western half of the Atlantic Ocean (its border is marked by the Mid-Atlantic Ridge), and the entire South American Continent, it forms a subduction zone running from the latitude of Panama to a point south of Cape Horn. The reader will remember this line of crustal weakness as the Peru-Chile Trench.

It has already been noted elsewhere that where sections of the earth's crust (plates) collide *in the vicinity of trenches*, those lining the floors of the oceans are invariably driven downward and over-ridden by the much thicker, heavier, and older sections of the continental land masses, for every trench marks a subduction zone. Thus it is with the Nazca Plate in this area. Buckling under the enormous weight and pressure of the overriding American Plate, its eastern edge has over millions of years been slowly sucked down into the depths of the Peru-Chile Trench, a movement creating stupendous subterranean pressures. These pressures pulverize some crustal rocks, forcing others downward into the hot material of the mantle. But such rock can be driven toward the surface, too. The enormous Cordillera de Los Andes was formed that way, very young mountains still in the process of creation, the longest range on earth that is above water.

Thus far, we have described certain facets of plate tectonics that are common to all plates. Wherein then lies the special interest attaching to the Nazca Plate? The answer is in its extreme instability. No earth zone of greater seismicity exists than that delimited by the Nazca Plate's eastern edge (west coast of South America), and to this must be added the secondary but still important factor of widespread vulcanicity. Geologically speaking, very few parts of this entire littoral could be considered "safe" places to live. The whole region is subject to natural cataclysm, particularly earthquake, and its hardy inhabitants have periodically fallen victim to some of the worst disasters in history.

As has already been noted, the records of strike by tsunami against South America's Pacific coast go back for centuries, and since the days of the Spanish Conquest have been well-documented. Thus, we find that this region was inundated three times in the latter half of the sixteenth century—in 1562, 1570, and 1575; twice in the seventeenth—1604 and 1657; twice in the eighteenth—1730 and 1751; four times in the nineteenth—1819, 1835, 1868, and 1877; and, to date, three times in the present century—1906, 1922, and 1960. Over the generations, Colombia and Peru have been hit repeatedly. However, Chile is most often the victim in such cases, both of the devastating earthquakes which periodically level her towns and cities and slay her people, and of the earthquake-caused tsunami which ravage her coast and destroy her ports and shipping.

We have seen what can happen in this area. The waves that smashed Valdivia in 1575 and Concepción in 1751 were briefly mentioned. Those that drowned Arica in 1868 were dealt with at greater length. At this point it might be illuminating to cite yet another example, that of the cataclysmic upheaval of 1960, not only because tsunami (called *los maremotos* in Spanish-speaking countries) were propagated, but because they resulted from a crustal displacement whose scope was subsequently recognized to be nearly unprecedented in the memory of science.

Chile, especially in its southern half, is a country of almost incredible volcanic and seismic instability, numbering between the parallels of 36° and 47° south latitude (say, roughly, between Linares in the north and the Golfo de Penas in the south) no fewer than 15 *active* volcanoes. To this circumstance may be added other factors. For one thing, the frequency of seismic shocks would seem to indicate that the process of continent-building has produced greater crustal imbalance in this section of the Peru-Chile Trench than any other. For another, a glance at a map reveals that two of the largest Pacific fracture zones, the Challenger and Fernandez, lie for the most part within the indicated parallels.

From the standpoint of what happened to the country in 1960, by far the most important topographical features are its two longitudinal faults, the one many miles inland and visible in certain areas, the other under the sea off the coast, its existence inferred by meticulous and extensive seismic studies. The inland fault, known as the Reloncavi, runs north and south through that portion of eastern Chile called the Lakes District, roughly from the town of Llifen in the north to the latitude of Castro in the south, a distance of about 171 miles. A few miles south of Llifen it passes directly under the volcano Puyéhue, one of the active volcanoes mentioned above, of which more later.

Thanks primarily to the work of the American seismologist, Pierre Saint-Amand, the position of the crucial oceanic fault is almost equally well-pinpointed. In Santiago at the time, a visiting lecturer at the University of Chile, he utilized his experiences as an expert eyewitness to write what is assuredly one of the most technically precise accounts of the disaster, a report later published under the title, *"Los Terremotos de Mayo—Chile 1960."*[1] In this report, consolidating the findings of seismological field studies conducted by teams from the university and the Chilean Geological Survey, he determined from the epicentral distribution of the aftershocks that a much greater fault than the Reloncavi runs north and south along the continental shelf and lies on the landward side of the Peru-

---

[1] In *Technical Articles*, article no. 14 TP 2701, (China Lake, California: Michelson Laboratories, U.S. Naval Ordnance Test Station, August 1961).

Chile Trench. This unnamed fissure begins approximately in the latitude of the city of Concepción, passes directly through the Arauco Peninsula, and extends southward to at least 47° south latitude.

Let us return for a moment to the Nazca Plate. We know that its line of subduction, the point at which its eastern edge buckles and begins to "dive" under the overriding American Plate, is marked by that longest crevasse on earth, the Peru-Chile Trench. Already noted is the ages-long grinding of these two plates and the enormous subterranean pressures thus created. The "skin" of the earth is thin, the friction caused by this grinding of inestimable force and intensity. It is not remarkable then that the western edge of the American Plate should be torn by it, a section of that edge fractured in lines that run parallel to the source of the friction, lines the dimensions of which diminish with increasing distance from the trench. Admittedly, we theorize. The entire concept of plate tectonics is theoretical. But the dictates of logic would seem to offer strong evidence that what happened in south-central Chile in that fateful year, as in others, was a cataclysm directly related to the relentless and inevitable process of plate subduction.

On Saturday, May 21, at 6:02 A.M. (well before sunrise in these latitudes at this time of year), the series of quakes which were to lay waste a strip of land 200 miles wide and 375 miles long began. In the coastal town of Concepción the damage was heavy. Walls, chimneys, church belfries, many structures of stone, brick, and plaster came down, while hundreds of people who had had no warning were killed before they could leave their beds. Sixty miles inland in the town of Chillán it was the same story, and the settlements of the Arauco Peninsula (at the south end of which lay the epicenter) were also heavily shaken. This first shock had a magnitude of 7.75 on the Richter Scale. Though considered severe, it was dwarfed by what was to come.

Half an hour passed before the second shock, though when it struck it was still dark. The violence of the first temblor was equaled. Damage became more widespread. But the frightened population of the towns had already deserted their homes, crowding into the streets, parks and squares, and this factor saved many lives. Again, the source appeared to be the same epicentral area.

Thirty-three hours passed. Aftershocks had occurred all during the day on Saturday, and during the night. On Sunday, their intensity began to increase, growing markedly strong in the afternoon. Just before 3:00, a quake struck that was the strongest yet (magnitude 7.8), causing those whose homes were still standing to again pour into the streets. This proved a most providential exodus. At 3:11 P.M. came the type of cataclysmic earthquake which in other times and places has slaughtered people by the tens of thousands. Its magnitude was later calculated at 8.9,[2] meaning that its power was over *ten times* that of the first temblors. It was to ravage the habitations of man. Valdivia and Puerto Montt were almost leveled, unrecognizable, with the former 80 percent destroyed. All the coastal towns were devastated, so hard hit that even the houses which did not collapse were wrenched from their foundations, twisted and shattered beyond repair. On the island of Chiloé in the Gulf of Corcovado, ruin was general. The town of Castro lay all but obliterated.

Later, the extremely fortunate circumstance of the main shock striking on a Sunday during daylight was generally recognized. Of the 39 great earthquakes recorded worldwide since 1897, only two had equaled this one in magnitude.[3] Had it happened on a weekday, with schools, factories and places of business crammed with humanity, the toll of dead and injured would certainly have been vastly multiplied. (One thinks of 180,000 deaths in the quake in Kansu, China, in 1920; 140,000 in Kwanto, Japan, in 1923.) But the people, warned by the previous shocks, knowing their homes to be already largely uninhabitable, had remained out of doors, thus escaping the destruction caused by falling masonry. No exact count of their losses was ever made. The estimates of the Chilean Government show a variance of 5,000 to 10,000 dead, with the figures of foreign observers averaging close to 6,000. The more than 200 aftershocks lasted until June 22 (a prodigious length of time), and ten of them were greater than magnitude 7.

It was stated that the first and second shocks of May 21 appeared to have the same epicenter at the southernmost part of the Arauco Peninsula. This situation did not persist. In fact, as other quakes followed on successive days it became obvious that over *four dozen*

[2]The figure was computed by the Institut de Physique du Globe at Strasbourg, France.
[3]Colombia, January 31, 1906; Sanriku Coast (northeast Honshu, Japan), March 3, 1933.

epicenters had developed, both within the land mass and under the sea, extending throughout south-central Chile. The power of this incredibly violent wrenching was fantastic. Roads, including the famed Pan-American Highway, were heaved upward in places, downward in others, and smashed to bits. Ponderous steel rails were twisted like spaghetti. In the seaports piers were warped and crushed, ships capsized, warehouses collapsed. Inland, concentrated along a 130-mile stretch of the Lake Region from Calbuco in the south to Villarica in the north, were hundreds of landslides, with homes and people buried under many of them. Located roughly in the center of the slide area, the volcano Puyéhue, dormant since 1905, began to erupt about 48 hours after the main shock, its cloud rising to over 25,000 feet. In its flank a fissure more than a thousand feet long had opened. The eruption continued for more than six weeks.[4]

It might be supposed that all of the above constituted human loss and misery on a scale not easily exceeded, but to that portion of the population living in the coastal communities it was the least of it. In the northen part of the stricken area (primarily, the Arauco Peninsula) and on the island of Guafo (southwest of Isla Chiloé), the land was severely uplifted, with what effect on man-made structures the reader may readily imagine. Elsewhere, the convulsions of the earth resulted in gigantic and catastrophic subsidence. From a point 20 miles north of Puerto Saavedra southward to the latitude of Osorno (a distance of about 140 miles), the coast as far as 18 miles inland simply sank to depths ranging between 6 and 12 feet, a phenomenon that was duplicated along eight-tenths of Chiloé Island's east shore. This meant, in effect, that within a few seconds a strip of land totalling 300 miles in length and 5,000 square miles in area had settled drastically and was open to invasion by the sea.

The invasion came. The people of Chile had seen it happen before. Those who could made for high ground.

About fifteen minutes after the great shock that occurred at 3:11 P.M., Sunday, May 22, the sea rose in the first of several incoming

---

[4]Subsequently, rumors spread of the creation of new volcanoes at this time. They were untrue, nor did any pre-existing volcano other than Puyéhue erupt.

floods whose high water mark far exceeded the normal tidal range. During this time, in all the bays and estuaries in the region of the disturbance, the submarine action of the quake caused the water to become exceedingly turbid, stirring up a heavy chop, causing in places (according to people who were afloat at the time) the shoals of sand to undulate toward the surface, then repeatedly sink and rise again, like white whales who breach toward the sun only to sound once more.

This phenomenon of a very rapid incoming tide which crested at 12 to 15 feet above the normal high water mark was common to the entire region where coastal subsidence had taken place. Some ships and boats parted their mooring lines. In the towns built only a few feet above sea level there was flooding, and the damage and terror caused by the rising water added to the already existing destruction. Then the sea went out. As usually occurs in such cases, it retreated with enormous speed, with the spine-chilling hissing and sucking noise, unbelievably loud, which accompanies its own vast turbulence and its passage over the countless millions of water-tumbled stones lining the shore.

At numerous points along this 300 miles of coast a good many boats went out with it, some of them ships of considerable burden whose cables parted or whose anchors failed to hold under the terrific strain of that savagely abnormal ebb. Almost without exception they were destroyed. Some fishermen, like the groups at Isla Chiloé and Corral, had taken their families and put to sea before the first inundation (called by the inhabitants the "smooth wave") had flooded the land. As seen from the shore, their boats lifted on the first incoming crest which passed beneath their keels harmlessly. But at 4:20 P.M. the second wave could be seen, a 26-foot-high green wall which because of the gradual offshore slope never acquired monstrous proportions or slowed its speed to the freight train pace so often seen elsewhere. This tsunami, traveling at a fearsome 125 miles per hour, overwhelmed the bobbing lines of boats, which were engulfed on contact and never seen again. Then it hurled itself against the land, smashing dozens of towns and villages to matchwood.

Fortunately, either after the first flooding or the first ebb, most of the inhabitants had withdrawn into the high country. Even so, Corral lost 80 people, Queule and Toltén about 40 each, Puerto

Saavedra and Isla Chiloé an undertermined number. To the south there was disaster. Apparently, due to ignorance or inadequate warning, the villages clustered at the estuary of the Río Maulin remained partially occupied. The toll was heavy here. Estimates of the number drowned run from about 1,000 to 1,500 people, nearly all of whom might have been saved.

Far out to sea, the third wave appeared one hour after the second had struck. Its height was an estimated 35 feet, but its speed was computed at about half that of its predecessor. There wasn't much left for it to destroy. Earthquake damage was heavy and widespread. The havoc wrought by the previous wave was nearly total. The third wave struck. For twenty minutes its surging waters lay deep over the land. Then it too withdrew with great speed, sucking out to sea smashed boats, the additional debris of many houses and waterfront buildings, and the corpses of people and animals. The ships at Corral were destroyed this way. One of them was the *Canelos*, grounded half a mile out when the lightning ebbing of the second wave deposited it on a sand bar. Then the third wave hit the ship, sweeping it inland *against the flow* of the Calle-Calle River as though that swift current never existed, and beaching it a mile upstream.

The fourth wave was just a repetition of the third, though in some places it was even higher. No additional lives were lost. All afternoon the sea advanced and withdrew, piling wreckage as much as 2 miles inland of ruined Puerto Saavedra, flooding fields north of the river port of Valdivia, so utterly obliterating some towns and villages that only the lines of the streets remained to mark their graves. Queule was one of the latter, a community as totally erased as though it had been hit by a nuclear projectile, only a few of its battered trees remaining.

And that was it so far as Chile was concerned. As difficult as it may be to believe, *five provinces had subsided*, the land settling to such a degree that serious flooding would have resulted even if there had been no tsunami. Later, observers flying over the stricken land in helicopters and slow-flying fixed-wing aircraft were exposed to sights they would never forget: roads and railroads that led nowhere, simply disappearing into what had become vast lakes or newly created, very extensive arms of the encroaching sea; entire communities wiped from the map, only water to be seen where once they had stood; many cities and towns and scores of villages laid waste, scarcely recognizable as former places of human habitation; farms devastated, in many cases simply gone, drowned; long sections of land uplifted, weirdly broken, smashed and deserted houses strewn on their slopes; once-thriving seaports turned into shambles, their docks and warehouses flattened, their boats and ships sunk, beached, or reduced to derelict hulks; great landslides covering and blocking all roads in the Lake Region, severing communications; the volcano Puyéhue growling in its furnace, fissuring its flank and creating steaming fumeroles—with what portent for the future behavior of the Reloncavi Fault only time would reveal.

There remains to be noted the effects of the Chilean tsunami in other parts of the world. The reader will recall that seismic sea waves move out in concentric circles, radiating from the center of the disturbance that caused them, be it a submarine landslide, volcano, or earthquake. Due to the configuration of the west coasts of North and South America, it might be supposed that such a wave originating in Chilean waters would close on California at an angle too acute to cause much material destruction. Yet, in rapid succession, San Diego and Los Angeles harbors were struck heavy blows, the combined damage to dock facilities and small craft totaling over $1 million, though no lives were lost.

As has been mentioned earlier, with regard to human casualties, in Hawaii it was a different story. At 3:11 P.M. in the Chilean time-reference it is 7:11 P.M. Greenwich Meridian Time or, in the military designation, 1911 hours. This, be it noted, was the time of the main earthquake shock. Shortly thereafter, reports from far-flung seismograph stations began to be radioed in to the Seismic Sea-Wave Warning System (SSWWS) Control Center at Ewa Beach, Oahu.[5] Information was logged and correlated. The record shows that by 2159 hours the computations had pinpointed the epicenter, by which time it was realized that the magnitude of the shock made the propagation of tsunami a distinct possibility.[6] The control center

---

[5]For details on the operation of the SSWWS, now the PTWC, see Chapter 12.

[6]As early as 2015 hours, the Government Hydrography Office at Valparaiso had correlated tide-gauge reports and noted a significant rise in sea level. However, for reasons never made entirely clear, their warning to Honolulu was delayed.

released a first tsunami warning to the island residents, meanwhile relaying all available data to other stations within the system. The tsunami were traveling across the Pacific at an average speed of 442 miles per hour. They could be expected to reach the Hawaiian Islands some 14 hours and 56 minutes following the birth of the first of them off Chile.

At 10:04 P.M., Honolulu began to receive information concerning the heavy sea-wave damage along the Chilean coast, thus confirming their suspicions, and issued a second, more positive warning. All night the technicians remained at their posts. At 4:45 A.M., they broadcast the expected arrival times of waves at various locales within the Pacific Basin, along with a final warning. How accurate were they? Their prediction stated that the first wave would hit Hilo at 9:57 A.M. It came in *one minute late.*

Already related was the tragedy of Hilo; the failure to heed the alarms, the foolhardy sightseers who went down to the waterfront to watch the wave come in, the resultant increased casualty figure of 61 dead and 282 injured. The waves had to go much farther to reach the coasts of Japan's northern islands. There, because the Japanese knew no precedent for damage from Chilean shocks, the agencies whose duty it was to warn the people were caught partially by surprise. More than 22 hours after its generation some 10,600 miles distant (8 hours after striking Hawaii), the first wave came roaring in against the shores and harbors of Hokkaido and Honshu, its height varying from 12 to 20 feet. Small, you say? Fishing boats weighing many tons were lifted over crushed docks and propelled into the business districts of such coastal towns as Shiogama and Ofunato (both on Honshu's Sanriku Coast) like so many toys, where their hulls proved irresistible battering-rams to buildings large and small. Port facilities were destroyed. Numerous craft of varying tonnage were sunk or rendered useless. The damage to public structures alone was estimated at $70 million while losses in private property and shipping ran to a staggering $350 million. Altogether, 5,000 houses were obliterated, leaving 50,000 homeless. Add to this the loss of livelihood and the figure of those disastrously affected rises to 150,000. Between 180 and 190 people died.

Alaska Again

Movement along the Denali Fault System: March 27, 1964.

To the seismologist, a study of any of the earth's many fault systems can be an endlessly fascinating task, for added to the irresistible urge to increase scientific knowledge is the very practical goal of reducing human casualties resulting from future cataclysms. This goal is not unattainable. Its accomplishment draws nearer with each positive step in the science of earthquake prediction.

According to the theories of continental drift and sea-floor spreading, the earth's crust fractures in its thinnest and weakest parts under the oceans and along continental margins, on the one hand creating the globe-girdling fissure called the Mid-Oceanic Ridge (and less spectacular fractures), on the other the vast trenches into whose depths the great tectonic plates are gradually subducted. Today it is believed that this process causes Europe to move away from North America at the rate of one inch annually, and shifts certain parts of the Pacific Plate by as much as four inches in the same period. In any discussion of the seismicity of southeast Alaska, it is the latter movement with which we are chiefly concerned. Measurements taken along the San Andreas Fault since 1956 show that the eastern edge of the Pacific Plate moves northwestward at an average rate of about two inches a year. What does this cause to happen in the north? Where the plate abuts against the 2,000-mile-long undersea wall of the Aleutian Trench it is subducted along that line and over the millennia passes very gradually beneath the earth's crust. There (just as occurs where the Nazca Plate slips under South America through the Peru-Chile Trench), rock-

77

shattering friction is created, along with pulverizing subterranean pressure. In the ages-long course of geologic time this friction and pressure has produced its recurrent effect on the crustal masses of both the Aleutian Islands and the Alaskan mainland. One aspect of it is the Denali Fault System.

Dating back to Alaskan prehistory, *denali* is an Aleut word and means "home of the sun." Before the Americans renamed it, 20,320-foot-high Mount McKinley was called *Denali* by the Eskimos, and the word is still a popular one. In Anchorage, for example, there is a Denali Theater and a Denali Elementary School, and the fault system bearing this name is the most extensive in the southeastern part of the state.

To draw a comparison, California's San Andreas Fault System has several major branches. Just east of San Francisco, there is the Hayward Fault and the Calaveras; farther south, the Sierra Nevada Fault and the Garlock Fault thread the Owens Valley; in the southwest, the Elsinore Fault runs north to pass through the San Fernando Valley. Exceeded in size only by the San Andreas itself is the San Jacinto Fault. Rising in Mexico's Gulf of California, this large fracture passes through the Imperial Valley and flanks the Salton Sea, joining the San Andreas north of San Bernardino.

Southeast Alaska's fault system shares points of similarity with this. It, too, is distinguished by horizontal or transform faulting, meaning that its movements are lateral, any given face invariably shifting in the opposite direction from its counterpart. The Cook Inlet Fault rises near Tuxedni Bay in the inlet whose name it bears, splits the land beneath Kamishak Bay, then traverses the length of the Alaska Peninsula to again lose itself in the sea just off Unimak Island's south coast; the Lake Clark Fault begins in the western Chugach Mountains, flanks Anchorage on the north, proceeds southwest through the Lake Clark region, and extends ninety miles into Bristol Bay; the Seldovia Fault, its eastern extremity much closer to Anchorage, passes southwest through the Kenai Peninsula (close to the town named for it), bisects Kodiak Island, and extends into the Gulf of Alaska to the vicinity of the Shumagin Islands. Lastly, there is the Fairweather Fault, by far the longest of the entire system. Beginning just north of Prince William Sound, it runs eastward close to the coast, then curves southeastward on a line extending down at least to Glacier Bay and possibly well beyond the 58th parallel.[1]

When one begins to study these Alaskan faults on a geologic map, features they share in common quickly become evident. For one thing, parts of all four lie within a region of great seismic instability. For another, three of the four (all except the Cook Inlet Fault) move toward near-convergence in the vicinity of the epi-central area about to be examined.

Admittedly, the Denali System is only part of a gigantic complex of faults extending throughout the entire Alaska-Aleutian arc, a complex of such widespread and powerful seismicity that in 1957 movement in its western region resulted in the longest-sustained and one of the most violent tectonic convulsions known to science. From January to November of that year hundreds of quakes were recorded throughout the Aleutian Islands, covering an area 750 miles long, subtending a longitudinal arc of 17 degrees. Twenty-two of the shocks exceeded magnitude 6.75. One was greater than 8. Fortunately, the severity of Aleutian weather (and, presumably, the existence of numerous active volcanoes) assures that area a sparse population.

In the eastern reaches of this fault-lined arc, however, another situation prevails. There the climate is much more salubrious and volcanic activity considerably less evident, so much so that, like the Russian immigrants who preceded them, Americans have found the glacier-carved Kenai Peninsula an attractive place to live, and have settled its shores from the innermost arms of the Cook Inlet to the fjord-indented east coast of Prince William Sound. Thus, the peninsula's principal city, Anchorage, and such lesser towns as Cordova, Valdez, Chenega, Seward, and Seldovia all lie within reach of one or more of the faults mentioned above, or of fault-generated oceanic disturbance. So, too, does Kodiak Island to the southwest, and the modest seaport that bears its name.

---

[1]Movement along the southern part of the Fairweather Fault is responsible for the titanic avalanches which periodically thunder down from the range of the same name, creating the enormous waves which make adjacent Lituya Bay uninhabitable. See Chapter 9.

*　　*　　*

It was March 27, 1964, Good Friday, the day on which Christ died more than 19 centuries before. For so early in the spring it was a beautiful day, a crisp 28 degrees at sunrise, but giving promise of sufficient warmth to make the school holiday pleasant for children and the coming activities of Easter a memorable experience.

If, on this day, a team of geologists had happened to take tests along the fault zone in question (either with seismometers near the surface or strain meters deep underground), they quickly would have discovered that the serene appearance of the land was deceptive. In at least one segment of the zone, their instruments would have detected the slight but ominous movement of subsurface rock called *dilatancy* by seismologists, a process invariably occurring when major fault slippage is imminent, involving gradual rock expansion due to the opening of thousands of tiny pressure-caused fissures. This phenomenon would have produced a sharp increase over the usual number of microseisms, or minute earthquakes and, because of refraction against successively splitting surfaces, would have slowed the subterranean passage of seismic waves through the crust.

We may presume there were other detectable phenomena. Almost certainly, in some areas of the fault zone, inclinometers (sometimes called tiltmeters) would have registered slight changes in the gradients of certain slopes, while the employment of magnetometers would have revealed the telltale alterations in the earth's magnetic field. Even observant laymen might have noted certain signs. For one thing, it is likely that the water level in reservoirs and wells in or near the fault zone would have undergone measurable changes. For another, animal behavior would have been abnormal. Domesticated farm animals would have become obviously restless, refusing food and displaying every sign of extreme nervousness, while many of those in the wild state would have exhibited symptoms entirely contrary to their natures.

An example of the latter did actually occur and was observed. The brown bears of Kodiak Island, which, with the closely related Montague Island species, constitute the largest and most powerful land predators on earth, left their caves two weeks earlier than normal, an exodus that occurred the day before the earthquake. An examination of their spoor proved revealing. Bears just out of hibernation are usually very hungry, their first act on leaving the den being to forage for food. These animals had no interest in eating. Their tracks made it clear they headed directly away from the fault area, and not at their usual lumbering walk or even at a lope—but at top speed.[2]

The day that had opened with such promise moved toward its close. In Anchorage, the most heavily populated (104,000) and economically important city in Alaska, a few last-minute shoppers were still in the streets, the going-home traffic light. This was the picture at 5:30 P.M. During those fateful moments, it is doubtful that many gave thought to Saint Matthew's Gospel account of the first Good Friday. "And, behold . . . the earth did quake, and the rocks rent." Still less were they thinking of what might be happening at a point within the earth's crust 150 miles to the southeast. There, close to those northern arms of Prince William Sound whose fingerlike extremities penetrated the Chugach Mountains, twelve miles beneath the Great Columbia Glacier one or more of the ancient faults forming the Danali System was in motion.[3] No one knows how long the pressure had been building in this region. Like some titanic spring drawing tighter and tighter, the stored energy had doubled and tripled and doubled again, confined by its counterpressure of trillions of tons of surrounding rock, until, at 5:36 precisely, the critical point was reached. Within the next six minutes the energy released was on the order of 200,000 megatons, the equivalent of 200 million tons of nitroglycerin, 12,000 times the force of the bomb-drop at Hiroshima. The shock waves rolled. They were to reach a magnitude of 8.5, the highest recorded on the North American continent since the Alaskan quake of 1899.

Anchorage was hard hit. In the downtown district, streets were lifted up, then cracked like whips, whole sections of them dropping as much as 30 feet and their buildings with them. Fourth Avenue was the most devastated. As perhaps nowhere else, the capriciousness of the temblor was revealed here, that incredible (though not inexplicable) selectivity so often seen in earthquakes and tornadoes.

---

[2]It was also noted that herds of domestic cattle moved to high pasture the day before the shock, several days earlier than usual.

[3]Judging from the location of the epicenter of the main shock, it is probable that the Fairweather Fault was involved. Considering the nature and extent of the gross block movement to the southwest of this area, involvement of one or more of the other three faults above named is equally likely.

On the north side of the street, the older side, was the Denali Theater, a scattering of bars and cafés, and a few shops and amusement parlors. On the south side was the bus depot, a number of stores, fur emporiums, and various business offices. The shock traversed the length of Fourth Avenue in a fraction of a second. As it did so, its high-frequency vibration liquified much of the alluvial detritus (most of which was clay) and loosely compacted fill under the street's north side, causing the ground to flow out from beneath a score of buildings. They simply sank down. Typical was the Denali Theater, an edifice which came to rest 10 feet below the level of the sidewalk, its marquee hanging only inches above the broken pavement. Structures on the south side remained virtually intact.

The quake struck many heavy blows. At Anchorage International Airport the 60-foot-high control tower came slamming down, killing the air traffic controller. At the Government Hill area flanking Elemendorf Air Force Base, landslides caused severe damage, destroying a hospital, homes, and installations, tearing apart the unoccupied elementary school. Such occurrences, however, were scarcely more catastrophic than the devastation in Turnagain Heights. Located just south of Anchorage proper, the city's most exclusive residential suburb consisted of about 300 homes, many perched rather precariously on a broad bluff overlooking the ice-jammed Turnagain Arm of the Cook Inlet. Like that part of Fourth Avenue already described, this bluff simply dissolved, carrying many broken homes with it into the frigid water, completely altering the shape of the coastline.

Admittedly, all this is tragically impressive. Yet, although property damage in Anchorage and its environs was great, the people there were spared the ultimate terror, protected from it by height (most of the city stands about 100 feet above sea level), by the intervening bulk of the Kenai Peninsula, by the simple fact that the quake's epicenter lay on its opposite side. Not so with some of the other towns. Because of the shock's location, settlements flanking Prince William Sound were open to devastation from the sea. So were those in the southern reaches of the Kenai Peninsula and on the east coast of Kodiak Island. And tsunami had been generated. Even while the motes of debris-stirred dust still hung in the air over stricken Anchorage, the great destroying waves were on their way.

*　　*　　*

To better understand why seismic sea waves were created in this instance, it might be well to briefly examine certain geological factors. The quake's initial shock produced a shallow "S" configuration in a strip of land about 310 miles long by 125 miles wide. This crustal deformation was such that the entire block of land to the northwest of the axis of shear subsided by an average of 6.5 feet, while that to the southeast of it was similarly uplifted.[4] The awesome power unleashed by the ground motion was most apparent when the result was viewed from the air. Around the shores of frozen lakes the ice was seen to be splintered; in some instances, strong water oscillation had cracked it from one bank to the other.[5] Throughout the northern half of the Cook Inlet it was also shattered, and in most of the bays along the deeply serrated coastline from the eastern reaches of Prince William Sound to the western extremity of the Kenai Peninsula. There were appalling landslides, particularly along the seacoast. Glaciers were split, *Bergschrund*-size[6] crevasses yawning where none had been seen before, while in the higher mountains thundering avalanches had sheared huge sections of forest away, leaving the lower slopes denuded of vegetation.

During the quake, the axis of tilt ran southwestward almost through the center of Prince William Sound, a motion generating a violent counterflow of water toward the northeast. This tumultuous sea, rushing into the many fjords bordering the sound, reached record run-up heights, then, reversing itself when balance was attained, swept back in the opposite direction. It was only the preliminary action. Actually, the area affected extended beyond Kodiak to the Trinity Islands; it was split all the way; the glacier-carved expanse of the Kenai-Kodiak spine was depressed; the submerged region to the south of it was uplifted. As a result, it has been estimated that more than an additional 6,200 cubic miles of water

---

[4]The greatest uplift (about 33 feet) was in the Montague Island area.

[5]See the section on seiches in Chapter 9.

[6]*Bergschrund*. German. Literally "mountain crevice." A term used in Germany, Austria, and the German cantons of Switzerland to denote only the largest of crevasses whether of rock or ice.

temporarily swelled the sea covering the shallow continental shelf, a condition making the propagation of tsunami inevitable.

Kodiak. The island of Kodiak, now valued at over $90 million, is approximately equal in area to Connecticut. Separated from the Alaska Peninsula by the turbulent waters of the Shelikof Strait, its fjord-fringed northwest coast faces the volcano-studded wilderness of the Katmai National Monument across the strait, while its only town of any size, the port of Kodiak, fronts on the Gulf of Alaska.

Looking at this town today, noting the health and activity of its bustling fish and canning industries (based primarily on the king crab), its paved and neatly laid-out streets, its attractively modern buildings, one tends to forget that little more than a decade ago it was a somewhat primitive settlement whose only claim to fame was the rather nebulous one of having been for thirteen years the administrative capital of Russian Alaska.[7] It is as though what happened to this community in 1964 rekindled in its population a kind of special stubbornness, a determination not only to erase the marks of widespread ruin but to build a town superior to its predecessor, bigger, more sophisticated, more attractive. It was perhaps characteristic of a people who, having seen their parents and grandparents evacuate in 1912 to avoid erupting Mount Katmai's threat of annihilation, well aware of the region's hazards, nonetheless returned to rebuild their shattered town. In 1964, such courage proved common to many Alaskans.

At the time of the quake, the island had an estimated population of 7,500, about 2,600 of whom resided in Kodiak City. In order of commercial importance, fishing, ranching, and hunting were the chief means of earning a livelihood, while some of the islanders worked as maintenance personnel at the Kodiak Naval Station. For the Aleut fishermen who formed the majority of the citizenry it was a largely uneventful, almost idyllic life. Their Kodiak Island villages of Old Harbor and Kaguyak had remained virtually unchanged for generations, as had their settlement on nearby Afognak Island. As

[7]Vestiges of this vanished civilization may still be seen today in the Baranof House Museum (once the home of Alexander Baranof, Russia's empire builder), and in the Russian Orthodox church with its blue Byzantine domes.

in big, important Anchorage that day, no one thought of disaster.

Moving at its tremendous speed of several miles per second, the primary surface wave of the great quake struck Kodiak only moments after rocking Anchorage, its low-pitched rumbling noise of tearing rock rising in swelling crescendo. The *Kodiak Mirror*'s editor, Karl Armstrong, described his attempts to walk in the midst of the frightful shaking as "like marching across a field of Jell-O." He managed to reach a telephone. "But when I tried to get Anchorage," he later reported, "the line was blocked. I knew then that the quake was everywhere."

Fortunately for the inhabitants, their homes were more than 200 miles to the southwest of the Great Columbia Glacier beneath whose ice field lay the temblor's focus. Because of this, there was time for a tsunami warning from the Fleet Weather Station to be received at the U.S. Naval Base and by residents of Kodiak City.[8] Some heeded it immediately and headed for the hills. Many others made the time-honored mistake of ignoring it, deciding to wait for further information. Only an unusual stroke of luck could now save them.

Down in the bay, the scene that presented itself at close of day had a strange dreamlike quality. On the darkening water all of Kodiak's 160-odd crab and salmon boats were fighting their moorings, hulls rocking and jerking, masts snapping back and forth in the characteristically erratic fashion of harbor-bound craft in a mighty earthquake. One of them was the *Selief*, an 86-foot, barge-hulled crab boat displacing 131 tons. At supper in the galley, Captain Bill Cuthbert could scarcely believe what was happening. Were it just a single blow, the impact of being rammed might feel like this, but the heavy buffeting was repeated and continuous, no way to account for it. He staggered out on deck. In the gathering darkness the scene that met his eyes was unbelievable. No boats in the world would behave like those wildly pitching craft unless— suddenly, he knew what it was. He switched on his radio and tuned in the standby emergency frequency, quickly picking up the voice of state trooper Don Church. The earthquake was confirmed. A few minutes later, a tsunami warning was broadcast, then repeated over and over.

[8]The Fleet Weather Station, Alaskan Sea Frontier, is an outpost of the Alaska Tsunami Warning System, operational in 1967. See Chapter 11.

As with any skipper in such a situation, Cuthbert's first thought was to weigh anchor at once, to clear the harbor with all possible speed and try to gain sea room while there still might be time. Then he remembered his engine was inoperable. He had a choice. His boat was moored fast to the dock. He could go ashore at once and climb into the hills with the others, or stay aboard and take his chances. But his boat was his work, and his work was his life. He went back to the galley and poured more coffee. Showing equal grit, his two crew members joined him.

The first sea wave arrived at 6:47 P.M., more than an hour after the initial shock. Exactly as occurred in Chile in 1960 (and in many another region), in the now dark harbor it took the form of a smooth but rapid rising of the sea, never cresting, therefore unable to break. It simply came in, causing some flooding along the waterfront, but little damage, a silent and eerie harbinger of what was to come. It served to sound a general alarm. Those who had ignored the weather station's warning could not ignore this. With blankets, flashlights, and quickly gathered food, they made for Pillar Mountain and climbed toward high ground, an exodus that saved a great many.

Then the tide went out. It was loud in its retreat, creating that weird, half-metallic sucking noise which identified its true nature. The fishing fleet bottomed. There in the darkness, *Selief* settled on the mud like the others, her beamy, flat-bottomed hull keeping her upright. The sucking noises died away.

Cuthbert knew what was coming. He was mariner enough for that. As it happened, he didn't have long to wait. Within a few minutes the second wave appeared, but very unlike the first. It was 30 feet high when it roared up the channel. As it struck each boat successively, like an enormous, upthrusting battering-ram, it hurled them ahead of itself, propelling their tons of dead weight over the stone jetty like missiles shot from a gun. Ponderous wooden piers vanished instantly; lines parted; heavy steel anchor chains snapped. In the near island channel, deserted docks and canneries belonging to the Alaska King Crab Corporation disintegrated, their debris swirling inland to add to the smashed remains of dock facilities serving the U.S. Fish and Wildlife Service and the Standard Oil Company.

As often happens when nature goes berserk, there were freakish occurrences. Berthed side by side at their slip were the graceful-hulled, well-found fishing boats, *Lucky Star* and *Victory Maid*. Later, the *Victory Maid* was discovered in mid-bay, capsized, heeled over on her portside beam ends but still afloat, supported by a ripped-off section of dock, pilings and all. She had been the inboard boat. The outboard craft, the slightly smaller *Lucky Star*, was never seen again.

The *Selief* proved her qualities. Ugly as she was, with lines as ungainly as those of a garbage scow, she nonetheless remained afloat. Like other boats, she was swept into town by the second wave, her heavy-timbered hull smashing through the walls and roofs of buildings as though they were cardboard. In the darkness of black water and churning debris, Cuthbert had no way of knowing where he was. When the wave receded, the cessation of motion told him he was "high and dry." All three men were still alive. They sat tight and waited.

Fifty-five minutes passed before the third wave struck. Reportedly, it was higher but lacked the vicious speed of the second. *Selief* took off again. More structures fell before her crunching impact. When the wave went out, she lay atop the smashed remains of what had been a store of some kind, her square-bilged hull apparently still intact.

Cuthbert and his crewmen waited. When the fourth wave came in they were ready for it. It carried *Selief* still farther into town, depositing her on a wreckage-strewn slope. Somehow, they managed to get a line around a telephone pole. Hour after hour they rode it out, finding themselves afloat at about 55-minute intervals, noting that each wave lacked the force of its predecessor, that none after the third one was really dangerous. About 3:00 A.M. the situation stabilized. Boat and crew had survived the ordeal.

Cuthbert remembered the aftermath, how the marine operator finally reached him by radio. "When she got to me, she said, 'Where are you, Captain?'

"So I told her—'By dead reckoning, in the schoolhouse yard.' "

He was right, too. Around the grounded *Selief* were the broken timbers of what had been the Kodiak schoolhouse—five blocks from the waterfront.

Narrow escapes were many that night. Bill Cuthbert wasn't the only one to display a sense of humor. There was the crab fisherman whose anchor found good holding after the worst wave had struck, but who recalled the awesome power of its debris-laden backwash. He remarked on it to the National Geographic Society's William Graves.

"I didn't mind the little stuff," he said, "but then I got hit by the Standard Oil Company building. Lousy thing, she cracked my bow."

Afterward, some of the citizens told of the "voyages" of Kraft's General Store. The ebbing of the first wave took it over the jetty and out into the bay, according to eyewitnesses. The second wave brought it back in, but again the ebb took it out. Apparently, it was the third wave that brought it to its final "mooring," battered but still whole, perhaps 200 yards from its original Benson Avenue site.

A crab boat captain recalled the scene for the same interviewer. "Couple of boats went over the jetty and come back without a scratch. But that's nothing. Kraft's store, she done it twice."

Perhaps typical of what happened to several native settlements along the coast is the account of how the sea dealt with Kaguyak, the Aleut fishing village previously mentioned. As with Kodiak, here also the first wave appeared as a suddenly rising flood tide, rapidly flowing in for a time, then reversing itself and ebbing. This, coupled with the quake itself, was enough to alert the people. Without hesitation, all 40 of them climbed an adjacent hill and waited. In the gathering darkness they saw the second wave push into town, though here, unlike its counterpart at Kodiak, it was neither large nor powerful. The villagers chattered excitedly. With night fallen, it would quickly get cold. They would need provisions: blankets, food, flashlights, radios to monitor broadcasts from Anchorage. Most thought the second wave would be the only one of any size, and that had already come and gone. Several men agreed to go for supplies.

One member of the party hadn't been so sure about the finality of the second wave. A leader among his people, he was a 26-year-old Aleut with the improbable Anglo-Saxon name of Roger Williams. He didn't argue with the others but he remembered having heard something as a child, a tale told by the elders. "The *third* wave is the one. Look out for the *third* wave."

Arriving at his house, Williams separated from the others, got the things he needed, then left the water-ravaged, but still largely intact village at a dead run, assuming his companions would do the same. But they hadn't followed him. It was only when he rejoined the refugees on the hill that he discovered the truth. He'd left them behind.

Nearly an hour passed. The six men still in town took a lengthy time, uncertain of the risk but intent on their task. Then the third wave rose out of the sea. It was big. They couldn't see it clearly but they could hear it. In desperate urgency they climbed into two fishing dories beached near the houses, having swung their bows round to meet the wave head-on. They waited. Maybe, they prayed.

No one knows what happened to one of the boats. Two of its occupants were never seen again. The third was found the following day, drowned, his body caught beneath a ripped-off house roof. Incredibly, the other three men survived. Swept through the doomed town on a rampaging, thunderous tide, they were carried near the point of the wave's highwater mark, then sucked seaward in its vicious backwash. They watched their chance. As the dory raced past the shoulder of a hill, they made a wild leap for tangled underbrush and hung on, a hundred-to-one shot that paid off.[9]

Although the loss of life in Kaguyak was small, the settlement itself was destroyed. When the waters receded for the last time, only a few shattered structures remained.

Valdez. Three hundred and five miles due east of Anchorage, founded at the turn of the century, Valdez lay at the head of one of the many fjords splitting the inland shores of Prince William Sound. Scenically, it was a marvelous location. On three sides could be observed the ice-covered reaches of the Chugach Mountains, spectacular enough to win for the region the title "The Switzerland of Alaska." On the fourth, at the seaward end of the Valdez Arm, was the water of the sound itself, a broad bay nearly landlocked by the barrier islands of Knight, Montague, and Hinchinbrook. Tourists were drawn to this locality. With its picturesque peak-roofed houses, wood-planked sidewalks, and "gingerbread"-trimmed façades, it somehow evoked the shades of the old days of the Gold Rush, of hard-

[9]Even the character in Poe's short fiction tale "A Descent into the Maelstrom" could hardly have undergone a more hair-raising experience.

bitten miners, teams of malamute sled-dogs, and *fancy* dancehall women. In 1964, it had two disadvantages. It lay within 22 miles of the Great Columbia Glacier. Most of it was built on detrital material, the disintegrated rock of a receding glacier's terminal moraine.

There were 1,100 people living in Valdez, although 100 or so were listed as "transient workers." Many were Aleut. The majority were crossbred stock, the descendants of Indians who had intermarried with the original Russian colonists. Good Friday was a lazy day. School was out. Most of the stores were closed. Yet, holiday or not, someone had to unload supplies from the ship now in port, the Alaska Steamship Company's 400-foot, 10,000-ton freighter, S.S. *Chena*. Down at the dock, the crewmen and longshoremen worked at a leisurely pace. Some exchanged good-humored banter with the people who stood watching them, a small group containing women and children. The sun began to go down. At six o'clock they were due to knock off for the day, after which the pier would be almost empty. It was 5:30 P.M.

Suddenly, the quake struck. Aboard *Chena*, Bernard Whalen and his friend Jim Aubert felt the ship shake to her keelson. Whalen thought the crane-like "jumbo gear" had carried away on the foredeck. Aubert, close to number 3 hatch, later recalled he believed the ship was getting under way. Docker Bob Hardin was down in the hold. He and his partner had just changed positions as driver of their cargo-raising forklift. Surrounded by the trembling steel of the hull, Hardin dashed for the companionway, only to be struck by beams falling from the hatchway above. When he finally reached topside his eyes beheld a bizarre scene. Forward, a dock shed roof had been hurled across the wildly pitching *Chena*'s deck. Ashore, fuel oil tanks, once high and dry, now jutted crazily from racing black water that seemed to have gone mad. In the confusion, he tried to see Valdez and couldn't find it. Then he realized that his line of sight was disrupted by the crest of a fast-moving wave, a wall of water rising between the ship and the land, rising to *break* over Valdez.

It was this wave that plunged so many into mourning, this, far more than the quake itself, which the survivors in Valdez were to remember with horror. On the pier were 28 men, women, and children, including stevedores and crew members. They saw the sea rising. Some stood paralyzed with fear. Others made a mad dash down the 300-foot-long dock toward the beach. The effort was useless. According to observers aboard *Chena*, those in the lead had covered less than a third of the distance when the water roared over them. The pier was ripped apart instantly, then submerged; planking, railings, barnacle-encrusted pilings and people—all smothered and dispersed in the terrible maw of that fiercely roiling current. By the time Hardin saw the wave it had passed the ship, and the people and the pier were gone. But Aubert had seen it from the deck.

Speaking of the doomed structure, he later reported: "One second she was there, the next she just wasn't. I saw maybe a dozen people turn and break for the beach, but she was a long pier—maybe a hundred yards—and they hadn't the littlest chance. She was sucked under all at once, like—well, just like the bobber on a fish line when a big one hits. My eyes can't seem to get rid of it."[10]

Ashore, ruptured fuel oil tanks now ignited, setting the waterfront ablaze, while *Chena*'s captain and crew fought doggedly for their lives. This ship's brief saga of stupendous struggle is probably one of the most singular in maritime history. During the course of this thundering nightmare she bottomed at least three times (some crewmen swore it was four), smashing down into the mud of a harbor which had originally been dredged to a depth of 35 feet. The first such drop sheared every line aboard. Theoretically, the impact should have started her seams and opened her to the sea, but it didn't. As she lurched sickeningly, injured men screamed in agony, but the moment the wave's backwash put him afloat again the captain, superbly cool, rang down to the engine room for power— and got it. The engines suddenly throbbed with renewed life. The blades of the propellers bit into that savage tide, their reverse thrust giving the ship sternway as she swung clear of the land.

At times, the raging sea had its way with her. From the bleak elevation of Thompson's Pass, soaked and freezing refugees recalled seeing her above the rooftops, her spinning propellers out of water. She crushed a warehouse, burst through it like a knife through

[10]Interview with William Graves.

cellophane, then heeled over nearly on her beam-ends. At that moment the risk of foundering was paramount, but skipper and crew, aided by phenomenal luck, met the challenge.

Propellers out of water means lack of control, an exceedingly perilous moment when a sea striking from abeam can cause a vessel to broach-to, suffering instant capsizement. But *Chena* righted herself. Again and again she yawed crazily. Always the skipper restored her to an even keel, fighting to maintain headway, desperately trying to avoid the debris of wrecked buildings and boats, and jagged masses of broken shore ice. At last he stood down the fjord, heading for open water. Ultimately, his battered vessel reached Prince William Sound, a ship that should not have lived, but did.[11]

Seward. If Montague Strait can be taken as part of Prince William Sound, and it is, then the town of Seward may be said to lie some 52 miles due west of the sound, at a point where the south coast of the Kenai Peninsula becomes a virtually unbroken series of deeply serrated, glacier-carved fjords. Facing the open seas of the Gulf of Alaska, specifically, Seward lies at the head of narrow Resurrection Bay, and from the ravages of ordinary storm waves finds adequate protection behind the peninsula of the same name.

Prior to 1964, Seward had several claims to a localized fame. Economically, its importance belied its small size (population 1,700), for as chief southern railhead of the Alaska Railroad, it was the principal transfer depot for petroleum and other products destined for the interior, and benefitted by a direct route to Anchorage via the Seward Highway. It was a tourist and sports mecca. As one of the oldest towns on the Kenai Peninsula, it had an interesting and colorful history. Its recreational facilities were varied and popular, its annual Silver Salmon Derby well known. So industrious and civic-minded were its citizens, so successful their attempts to improve their community in every way, the town was scheduled to receive an All-American Cities Award for civic improvement—but the quake struck a week prior to the date of conferral.

In the geological sense, the entire waterfront area was as unstable

---

[11]*Chena* made for Whittier at the Sound's northwestern extremity. There she was moored amid the wreckage of the dock owned by the Alaska Railroad.

and disaster-prone as it could possibly have been. Like that at Valdez, the off-shore slope was steep, varying between 30 and 35 degrees, and here also the foundations of any sea-fronted structures were laid in waterlogged alluvial detritus.

Nonetheless, three important companies chose to build on this site: the Alaska Railroad, and the Standard and Texaco oil companies. The railroad operated the two docks; with their adjacent warehouses they were worth over $4 million. The oil companies owned 16 oil storage tanks, 8 apiece, big, modern installations specially designed to resist the ravages of inclement weather.

Sundown approached. The hands of the clock stood at 5:36. As it did in so many other places, the ground suddenly shook and split open, shifting subterranean pressures sending geysers of water skyward. As their buildings swayed and their streets undulated, the citizens of Seward knew it was more than the usual sort of temblor, but had little time to think about it. Half a minute after the first shock, to the north of Washington Street, 4,500 feet of waterfront lurched into the bay. Warehouses, piers, and loading cranes went with it. So did all eight of the Standard Oil Company's big tanks. Other structures followed. The small boat harbor disappeared. Everything north of San Juan dock disintegrated, headed for the sea bottom. The tanks owned by Texaco stood on firmer ground. It didn't save them. Within seconds, most of them exploded, scattering burning oil for hundreds of yards. Then, with the same dire suddenness as at Valdez, allowing no time for a warning, the first tsunami came racing up the bay.

As to the height of this wave there seems to be no exact consensus, though the majority opinion places it at about 20 feet. It was bad. It did heavy though not pulverizing damage in the dock area, sank some boats, carried burning oil through the waterfront district and started some fires. It receded with the usual effect. The debris of broken structures was carried with it, the speed and power of the ebb (together with the impact of the wreckage it bore) proving sufficient to smash some installations which had survived the initial onslaught. Seward caught its breath. A handful of people, mostly longshoremen and railroad workers, reentered the dock area for salvage purposes. As the ground trembled periodically with the aftershocks, the minutes passed. Dazedly, the people dug themselves out. Nearly an hour went by. Then, at the seaward end of tightly constricted Resurrection Bay, the second wave appeared.

About this wave there was little argument. Most surviving eye-witnesses agree it was approximately 30 feet in height. Engulfing the remaining small boats as it sped along, it smashed into the docks, instantly crushing piers, slips, and moored vessels, then hurled itself over the ice-streaked, sodden land. On its way it had picked up more burning oil, thousands of gallons of it. As it burst over the docks its crest was a raging inferno of flame moving at 100 miles an hour, a crest that shone in the night. Railroad boxcars were hurled through the air like straws in a hurricane. Ponderous electric locomotives were tossed end over end as though they were toys, while the heavy steel rails they had rested on were ripped out of their beds and twisted to pieces. Nothing stood against this wave. The town's waterfront was demolished. The surviving boat or building was the freakish exception.

Observers recall with awe a once-in-a-lifetime sight. The fury of the waves had not only shattered the docks, but broken off the thick wooden pilings which supported them. Because they were water-logged in their lower ends, they now floated upright. Tarred to begin with, then soaked with oil, most of them were afire. They spread across Resurrection Bay "like a string of candles," indeed like de-votional candles lighted to mark the grave of all that was left of Seward's once-thriving waterfront. Inland too was afire, for the oil-bearing tsunami had spread flame everywhere with their advance. There were six waves altogether, but in the years to come people would speak mainly of the second. It was that one which would cause their faces to blanch, their brows to darken in grim recol-lection.

For the purposes of our study of seismic sea waves, it is unnecessary to further detail the earthquake losses in Anchorage, a city that was spared their devastation.[12] Tsunami destroyed Kodiak, along with 46 crab and 55 salmon boats. Additionally, such villages as Kaguyak and Old Harbor on Kodiak Island, and Chenega ($2 million damage,

---

[12]In what can only be considered a providential miracle, out of a population of 55,000 the dead numbered only 9.

25 drowned out of a population of 72) were virtually wiped out.

Seldovia (population: 500) was located on the westernmost promontory of the Kenai Peninsula, facing the Cook Inlet. Warned by radio, the people climbed to high ground, only to return to a town where land subsidence had dropped the waterfront by more than 3 feet, creating partial submergence during high tides. The small boat harbor was ruined, heavy pilings and slips gone, craft of varying burden sunk, but Seldovia itself was spared extensive water damage. Its location saved it. Since it was on the opposite side of the peninsula from the source of the disturbance, the only waves that struck it were "backlash" oscillations rebounding from across the inlet.

The same was true of nearby Homer, a community of 1,200 people on the north side of Kachemak Bay. When the Homer Spit was flooded, the port facilities were rendered useless. Docks and canneries were heavily damaged and waterlogged. Many public and private structures were destroyed or would have to be razed. But, like Seldovia, Homer had been warned in time. It reported no human casualties.

Southeast of Anchorage, along the Turnagain Arm of the Cook Inlet, other small communities such as Girdwood, Portage, and Whittier were hard hit. But, although the latter suffered 13 deaths and rising water did heavy damage, no direct tsunami action occurred there.

In examining the settlements scattered around the shores of Prince William Sound, the observer was made increasingly aware of the importance of two factors. The first is that the Seismic Sea-Wave Warning System, as then constituted, was not capable of issuing an early alarm to coastal dwellers living in close proximity to an earthquake epicenter. There simply wasn't time. Near a place of habitation, the following scenario usually occurs: a fault shifts, shock waves spread in all directions, tsunami strike immediately and with fearful speed. This happened at Valdez, and even at more distant Seward, for between the two the time distinction for strike by tsunami was negligible. At the southeastern extremity of the sound, Cordova too was devastated ($34 million damage), but, again, a fortunate circumstance saved it from the worst effects of tsunami action. Against waves originating at the north end of Prince William

Sound, as these did, "the friendly city" opposes a barrier. It is the nearly unbroken line of the Hinchinbrook Island spine, the eastern extremity of which juts into Orca Bay.

The second factor of importance is that of the necessity of dispelling an old myth: namely, the belief that no earthquake except a submarine one can propagate a tsunami. In 1964 this was clearly disproved. The mighty Alaska quake of that year had its epicenter on *land*, with its focus at a point some 12 miles beneath the Great Columbia Glacier. The ice mass was heavily fractured by subterranean displacement. Throughout much of the area whose shores are washed by the waters of the Gulf of Alaska, the aftershocks continued for over 16 hours.

7

Japan, the Ever-Vulnerable

The Japan Trench and the Tuscarora Deep. The state of Japanese seismicity. Scourge of the Sanriku Coast.

While it is true to say that in any compendium of worldwide vulcanicity measurement the islands of Indonesia must clearly rank first, it is equally true that of all the island arcs of the Pacific the four islands comprising the homeland of Japan stand first in seismicity, with the largest of them, Honshu, being chief among the four.[1] Unfortunately for the Japanese people this scenically beautiful but seismically unstable piece of land contains the great majority of the nation's population, most of its major cities, and more than five-eighths of its total land area. Fifty-eight provinces are located on Honshu. Out of an aggregate of 47 prefectures in the nation, this one island contains 34.

To the geologist examining a map of the Pacific Ocean floor the reason for Japan's instability becomes obvious. If all the water were removed from the Pacific Basin it would be seen that a long series of vast trenches divides the Asian mainland (and the Japanese home islands) from the entire northwest quadrant of the Pacific; that, moreover, though these trenches running from north to south take the names Kuril, Japan, Izu, and Mariana, they are in reality all part of the same enormous trench system and have been separately named merely for geographical convenience.

[1] From north to south they are Hokkaido, Honshu, Shikoku, and Kyushu.

97

Let's take a look at this gigantic depression. At its southernmost point (i.e., the southern extremity of the Mariana Trench) lies the greatest ocean depth to be found on earth, the famed Challenger Deep, marked on the charts to show a depth below sea level of 36,198 feet.[2] From this position, moving north into the latitude of southern Japan (Kyushu Island), we come to a second deep within the same depression, the Tuscarora (32,800 feet), and have now arrived at the conjoining point of the Izu and Japan trenches. Though, to the northeast, the very active seismicity of the arc running from Hokkaido through the Kuril Islands to Kamchatka is well-established, so far as concerns the effect of oceanic crustal instability on Japan itself we need look no further than the Tuscarora Deep and the Japan Trench of which it is a part. Many submarine earthquakes have their epicenters on the Tuscarora's west slope. Of these, a significant percentage have produced several of those tsunami which throughout history have ravaged the Sanriku Coast.[3]

Some idea of the crustal instability of the Japanese Archipelago can be gained by a glance at the seismic record. Every year, more than 1,000 quakes of various amplitudes and intensities are recorded in the main islands; seldom a day passes when people somewhere in the nation cannot detect minor tremors without instruments. Between 1891 and 1952 there were eight major shocks. All were of magnitude 7.3 or higher. One, the Sanriku Earthquake of 1933, reached 8.9.[4] That of Kwanto, in 1923, virtually destroyed Tokyo and Yokohama, and killed 140,000 people.[5]

Why is Japan so subject to severe seismic shock? As has been

---

[2]On some maps, the Challenger Deep is identified by its Soviet Russian name of Vityaz Deep. On January 23, 1960, the Swiss oceanographer Jacques Piccard (using the bathyscaph *Trieste*, designed by his father, Auguste) made a dive into the Mariana Trench with the U.S. Navy's Lieutenant Don Walsh. Touching bottom a short distance to the west of the Challenger Deep, the depth they achieved (35,800 feet) is still the world record.

[3]Sanriku is defined as the coast of northeast Honshu; roughly, from Matsushima Bay to Cape Shiriya, an area beginning 200 miles northeast of Tokyo as the crow flies.

[4]In doing so, it equaled the all-time record established by the Colombian quake of 1906 and the Chilean quake of 1960.

[5]Since earthquakes were first recorded in man's history the casualty figure of this terrible shock was exceeded only twice, both times in China: viz., Shensi in 1556 (830,000 dead); Kansu in 1920 (180,000 dead).

indicated, earthquakes whose foci lie at great depth within the earth's interior are seldom dangerous to man. However, where island arcs are concerned (particularly in the Pacific), the approaches from the oceanic side are invariably split by multiple crustal fractures, many of them lying between the deep flanking trench and the islands in question, all theoretically caused by plate subduction. Seismic foci within such trench systems are shallow, remaining so throughout the epicontinental sea (the sea to landward of the arc), their increase in depth commensurate with their distance inland from the trench.[6] The Japanese islands are typical of such a structure. As with almost all island arcs, a shallow sea (the Sea of Japan) separates them from the continental land mass. The deepest seismic foci in the islands themselves are found at a depth of approximately 185 miles, though the majority are far more shallow. Yet 600 miles inland from the trench system they may run as deep as 500 miles.

Because of their location in an area of high seismicity and the crowded conditions which have long prevailed in their great cities, the Japanese have traditionally taken a strong interest in the study of earthquakes. As early as the fifth century A.D. they had already established a royal library for the purpose of keeping accurate records, and over the years trusted officials of succeeding *tenno*s and *shogun*s carefully maintained them.[7] By the time of the fall of the Shogunate in 1868, more than 2,000 earthquakes had been recorded in the homeland, 223 of them listed as catastrophes.[8] Placing the greater emphasis on those which generated seismic sea waves, the following should be enlightening.

According to the Tokyo University seismologist, Akitune Imamura, between the years 1596 and 1938 the Japanese islands have

---

[6]In 1923, working with marine gravimeters off the Lesser Sunda Islands, the Dutch geophysicist Vening-Meinesz revealed the gravitational anomalies of the trenches, thus laying the foundation for that "gravimetric prospecting" which was to greatly increase man's knowledge of them.

[7]*Tenno* is the Japanese word for emperor, *mikado* being an inaccurate western variant. The *shogun*s were warlords of the higher nobility, military provincial governors who usurped the imperial power from the twelfth to the nineteenth centuries. The Revolution of 1867–68 ended the Shogunate and reestablished the emperors with the Meiji Restoration, the beginning of modern Japan.

[8]For purposes of categorizing the greater shocks, it should be considered that any quake whose magnitude is 8 and above is cataclysmic in its epicentral area.

been struck no fewer than fifteen times by major tsunami. Subsequently came those of 1940, 1941, 1944, 1946, 1952, 1953, 1960, and 1964. Of this latter group, some, those originating close offshore, took a heavy toll of life. Others, born in Alaskan or South American waters thousands of miles distant, claimed far fewer casualties.

The further one delves back into history, the more sketchy and tenuous the record becomes, for while earthquakes were meticulously recorded, tsunami, long believed a manifestation of divine wrath, were accepted resignedly. Already mentioned was the Sanriku Coast, a prime target. It is known that in the year 869 a seismic sea wave took 1,000 lives somewhere in this area; another in 1293 killed 30,000. But in both instances all pertinent details are completely lacking. In 1611 an 80-foot wave entered Yamada Bay, killing 5,000 people, but further information on this disaster likewise remains unavailable. In 1703, there was another strike in the same general region. This one is said to have drowned 100,000. If such a figure is even roughly accurate, it must stand close to the top as one of the greatest killer waves of all time. But information on it is exceedingly sparse. The author has not been able to discover the precise date of its strike, its size, or even the exact stretch of the Sanriku Coast inundated. Here, as it so often does, legend has apparently encroached on history, the typical result being to obscure.

Moving on a little in time, we know that in 1717 a tsunami struck southwestern Honshu, far from Sanriku, roaring into the Strait of Kii from the open sea, sinking more than a thousand boats in Osaka Bay, devastating such cities and towns as Osaka, Kobe, Kakogawa, Nishinomiya and Sakai. Reliable reports on human casualties are lacking.

It is when we enter the nineteenth century that Japanese accounts of tsunami become both more accurate and illuminating. Because they are chiefly the result of earthquake action, any discussion of them might be profitably preceded by a glance at the seismic picture. The large island of Honshu itself is simply but one extruded node of that thoroughly fractured volcanic spine which collectively forms Japan. Both latitudinally and longitudinally it is faulted from end to end, but, like most such areas of the earth, some of its faults are deeper than others, more extensive, more dangerous.

One such fault splits the floor of the Gulf of Sagami (Sagami Nada), sometimes erroneously called Sagami Bay, and movement along this fissure caused the disastrous earthquake of December 23, 1854, a time when Japan's feudal state was drawing toward its close. To describe the resultant tsunami to us, we have not one but two surviving eyewitness accounts, both by officers aboard the Russian frigate *Diana*. Though there is general agreement, their reports differ in certain details (such as the time of the initial shock), but are not the less interesting for that. The first, by the officer of the watch, reads as follows:

We felt the first shock at 9:15 A.M.; it was very strong, and it went on for two or three minutes. At ten o'clock a huge wave rushed into the bay, and within a few minutes the whole town was under water; the many ships at anchor were thrown against one another and seriously damaged—we immediately saw a great deal of debris floating. At the end of five minutes the water in the bay swelled and began boiling up, as if thousands of springs had suddenly broken out; the water was mixed with mud, straw, and every kind of rubbish, and it hurled itself upon the town and the land to either side with shocking force: all the houses were wiped out. At 11:15 the frigate dragged her anchors and lost one of them; presently she lost the other and the ship was then whirled round and swept along with a strength that grew greater with the ever increasing speed of the water. At the same time thick clouds of vapor covered the site of the town and the air was filled with sulfurous exhalations.[9] The rise and fall of the water in this narrow bay was such that it caused several whirlpools, among which the frigate spun round with such force that everything on board was thrown down. Yet the frigate held her own in the midst of these gyrations; she turned clear round 43 times, but not without sustaining serious damage. Until noon the rising up and the falling of the water in the bay did not cease: the level varied from 8 to 40 feet in height. Towards two o'clock the bottom of the sea rose again, and so violently that several times the frigate was laid over and the anchor was seen in no more than 4 feet of water. At last the

[9]From the subsequent account it becomes clear that the town referred to was that of Simoda, a small port on the tip of the Izu Peninsula about 80 airline miles southwest of Tokyo. Today it is spelled *Shimoda*.

sea grew calm. The frigate floated in the middle of an inextricable tangle of her own rigging and twisted chains. The bay was nothing more than an expanse of ruins.

Obviously, since a vessel caught by a tsunami in closely confined waters is almost invariably doomed, *Diana* was stoutly built and in excellent trim, her officers and men extremely capable. Nonetheless, the second account makes it clear that not only was she rendered totally unseaworthy, but in all likelihood would never put to sea again, her mere survival being in itself a miracle. Here now is the second eyewitness account:

> On the 23rd of December 1854, at 9:45 A.M., the first shock waves of an earthquake were felt on board the Russian frigate *Diana* as she lay at anchor in the harbor of Simoda, not far from Jeddo [now known as Tokyo], Japan. Fifteen minutes afterwards a large wave was observed rolling into the harbor and the water on the beach to be rapidly rising. The town as seen from the frigate appeared to be sinking. This wave was followed by another and when the two receded which was at 10:15 A.M., there was not a house, save an unfinished temple, left standing in the village. These waves continued to come and go until 2:30 P.M., during which time the frigate was thrown on her beam end five times. A piece of keel 81 feet long was torn off, holes were knocked in her by striking the bottom, and she was reduced to a wreck. In the course of five minutes the water in the harbor fell, it is said from 23 to 3 feet, and the anchors of the ship were laid bare. There was a great loss of life; many houses washed into the sea, and many junks [were] carried up—one, two miles inland—and dashed to pieces on the shore. The bay was beautifully fine and no warning was given of the approaching convulsion. It was calm in the morning and the wind continued light all day.[10]

One does not wish to dwell on the result had *Diana* been underballasted or her gun-breechings parted and a cannon carried away while she lay heeled over. She remains one of the few ships in history to have survived a seismic sea wave.

The next great Japanese earthquake, the Mino-Owari, struck 37 years later, in 1891. Caused by movement along the extremely unstable Neo Fault, a 70-mile-long rift extending from Fukui to

---

[10]"Notes of a Russian Officer," *Nautical Magazine,* vol. 25, no. 2 (London, 1856), p. 97.

the outskirts of Nagoya, it set off one of the mightiest shocks in Japan's history and destroyed or heavily damaged 280,000 dwellings. Since it was an inland quake and did not produce tsunami, it is mentioned merely as additional evidence of seismic conditions in this region.[11]

Five years after the devastation of the provinces of Mino and Owari, in 1896, there rose out of the sea one of those frightful phenomena which, like the deadly Krakatoa waves, was destined to be long-remembered. The disturbance did not originate in the Tuscarora Deep, but some 93 miles off the central Sanriku Coast on the west slope of the Japan Trench at about 2,000 fathoms, east-southeast of the important seaport of Kamaishi. It was a big earthquake. Its effect on the sea was clear enough proof of that.[12] But at that distancefrom shore (and at such a depth) the shock was barely perceptible on land, scarcely strong enough to warn a people who were used to constant temblors. They paid it little heed. Few could have guessed its epicenter was beneath the ocean floor. Dr. Imamura has left us an account gleaned from the old records.[13]

This great Sanriku tsunami came on a festival day when the towns-people were enjoying a holiday. The shock, which was felt about seven o'clock in the evening, was of the slow, undulating type, not vigorous, but of long duration. Unmistakably, a great submarine earthquake had occurred some distance off the Sanriku Coast. Soon afterward several minor shocks were felt—all aftershocks of the main disturbance. Knowing nothing of distant earthquakes, the inhabitants were ignorant of the peril that was imminent. Twenty minutes after the first shock, the sea was seen to recede, while a little past eight in the evening, noises like that of a rainstorm were heard. The tsunami was now on them—a wall of water some tens of feet high— and the holiday revelers, before they realized the awful situation, were swept away and drowned. According to one of the survivors, the sea was in an unusual commotion, followed by noises like those

[11]The magnitude of the Mino-Owari quake was never determined. Although it fearfully ravaged both provinces, destroying Mino's capital city of Gifu, the casualty figures were amazingly low: 7,300 dead and 17,000 injured. In some places the throw (or vertical displacement) of the fault exceeded 13 feet.

[12]Its magnitude was not computed.

[13]Akitune Imamura, *Theoretical and Applied Seismology* (Tokyo: Maruzen Company, 1937).

of a squall. Except a few who were saved through means seemingly miraculous, of the entire population of the towns and villages that up to the moment of disaster were thriving with life, not one of the population or a vestige remained. On the other hand, the fishermen who at the time were some distance out at sea and had noticed nothing unusual were, on their way home the next morning, amazed to find the sea for miles strewn with house wreckage and floating corpses. They then realized for the first time the tragedy that had been enacted the night before.

For the reader who by now is familiar with the subject, a reminder of the irresistible power of so mighty a wave is perhaps unnecessary. The official Japanese government casualty figures gave the following totals: "10,617 houses swept away, 2,456 houses partly demolished, 27,122 persons killed, and 9,247 persons injured."[14]

As with all tsunami, the height and destructive force of the wave varied according to the topography of any particular segment of the coastline. Its greatest height was at Yoshihama where the records speak of a terrifying 94½ feet. At Kamaishi, where (considering the configuration of the bay) its height could scarcely have been much less, it drowned or crushed 72 percent of the population and destroyed 88 percent of both publicly and privately owned structures, a result which for sheer magnitude of casualties approached virtual annihilation. At the village of Toni, 78 percent of the people were drowned and 72 percent of all structures wiped out, and all along the area of strike the above-quoted percentages were typical.[15]

What can we say of such casualties as these, such a demoniacal propensity for slaughter? It is, of course, the old story. Take an uninhabited stretch of coast in one of the desolate regions of the world, let it be struck by tsunami even larger than this, and the awesome natural cataclysm is born and dies beyond the ken of

---

[14]Ibid.

[15]In a watercolor by artist Barry Evans, done in the Oriental artistic style, we observe a scene which is undoubtedly one of the most spectacular and terrifying depictions of a tsunami ever painted. Judging from the size of the wave towering enormously over the doomed Japanese town, the artist had in mind either the cataclysmic tsunami of 1293 (30,000 deaths), that of 1703 (100,000), or that of 1896 described above. So little has Japanese provincial architecture changed over the centuries, it would be difficult to pinpoint the era.

man. Not so in heavily populated Japan. Not so in the nineteenth century, still a period of worldwide ignorance regarding such esoteric matters. As shocking, indeed as unbelievable as it may sound, such a wave bursting upon New York City today could easily leave half a million bodies in its wake, far more than that if the factors of mass panic and stampede were to take their logical toll.

Before we move on, it might be instructive to reflect on one of the more salient features of this 1896 tsunami, a characteristic it held in common with most others throughout history and with its brethren today: It was not even observed by Japanese fishermen at sea, who were totally unaware of it until their return the next morning. How sadly ironic. Today, our knowledge of the subject tells us that the speeding killer must have passed beneath their keels only a few minutes before slamming against that unsuspecting coast. Voice transmission by radio was not yet possible. But even if it had been, and the boats so equipped, they could have sent no warning. The wave passed under them, its height perhaps 2 feet, its wavelength many tens of miles long, its speed at least 400 miles per hour. The small craft may have rolled and pitched unnaturally for a few moments, perhaps even yawed and fallen off.[16] But what difference would it have made? The passage of this harbinger of death was both silent and unknown.[17]

Nature is capricious and often unaccountably contradictory. In Alaska in 1964 we had the case of a terrestrial temblor which, because it occurred several miles inland beneath a glacier, should not have produced a tsunami—but did. In Japan in 1923 the sea floor was wrenched by such a cataclysm as, theoretically, should have generated one of the mightiest seismic sea waves of all time— but did not. The wave it created could be said to be average, some 30 feet in height. Yet, had the fantastic data on this shock been

[16]To *yaw* is to steer erratically, unable to hold a course. To *fall off* is to veer to leeward, the direction toward which the wind is blowing.

[17]An arresting watercolor is that by the Japanese nineteenth-century printmaker Hokusai, entitled "The Crest of the Great Wave off Kanagawa." In the clutch of the tsunami we see two doomed boats, while in the background rises Mount Fujiyama (more correctly: *Fuji-no-yama* or *Fuji-san*). Since Fuji rises to the left of the wave's advance and is quite close, we may assume the caption refers to Kanagawa Prefecture whose coast encloses much of the Gulf of Sagami, rather than to the town of Kanagawa on Tokyo Bay. This being the case, the wave depicted might easily be the same which destroyed Shimoda and damaged the frigate *Diana* in 1854.

available to geologists in advance, very few would have predicted anything less than a crushing oceanic disaster.

Already mentioned was the Gulf of Sagami. That broad expanse of water (actually the entrance to Tokyo Bay), 42 miles wide, is split by very deep fissures. Its seismicity was well known. Earthquakes of the past had often had their epicenters there. However, near the end of the present century's first quarter it was due for a prolonged period of dormancy, or so Japanese seismologists believed. Tokyo University's Fusakichi Omori was one of these. Having studied the shock of April 26, 1922, he predicted that the Tokyo area would remain quiescent for at least a century.

He was wrong. On September 1, 1923, at least one of the larger faults beneath Sagami Nada moved radically and disastrously. Following the P and S waves, the slower surface wave passed through Yokohama 50 miles away, then Tokyo, 70 miles distant, and in 30 seconds of incredible violence reached a magnitude of 8.2. These cities collapsed. Within half an hour after the main shock 136 fires raged in Tokyo, creating gigantic vacuums, providing their own inrushing winds (called *fire storms*) as major holocausts will. Within 56 hours 366,262 homes were destroyed, 71 percent of the total. It was even worse in Yokohama. Twelve hours after the initial shock, widespread fires had wiped out 60 percent of the city; by the time they were out it was completely gutted. More than 140,000 people died in this single locale, most of them by fire. An additional 103,000 were injured.

To put the situation succinctly, not only was the earthquake of cataclysmic magnitude, capable of destroying great cities, but some of its results ran contrary to all previously known tenets of natural law. When, afterward, triangulation figures were compiled as part of a general topographical survey to determine cause, it was found that the entire north coast of the Gulf of Sagami and the west shore of Tokyo Bay had slipped in the direction of east-southeast, actually moving the cities of Tokyo and Yokohama by as much as 15 feet in some areas.[18]

Such extensive crustal displacement was not without precedent,

---

[18]One freakish effect of fractural convergence on this occasion was seen in the fact that, despite the general eastward slippage, the island volcano of Ooshima shifted its position some 13 feet to the *north*.

but that could hardly be said for the rest of it. Soundings showed that the floor of the gulf itself had undergone an incredible metamorphosis. Occurring in immediate proximity, subsidences and uplifts were on a scale so unbelievable as to excite the puzzled interest of the government. Accordingly, the Japanese Navy assigned four of its ships to make a second survey, an extremely exhausting one totaling some 83,000 soundings. The earlier results were corroborated. The new soundings indicated depth increases averaging 300 to 600 feet since the year 1912, with a maximum vertical disparity of 1,300 feet.[19]

It is perhaps needless to add that such figures virtually defy belief, for their equal is not remotely to be found in geologic history. In the Alaskan or Yakutat earthquake of 1899 (magnitude 8.6), uplift and subsidence at Disenchantment Bay varied from 9 to 47 feet in 1¼ miles. In the Assam Earthquake in 1950 (magnitude 8.7), the throw of the Chedrang Fault varied from 4 to 35 feet over a distance of 12 miles. But 1,300 feet? Various explanations have been forwarded, everything from submarine slumping on a titanic scale to the abrupt compaction of very loose sediments lining the ocean bottom. None of them seem satisfactory. Considering the well-known Japanese reputation for accuracy in such matters, despite its seemingly impossible claim, we can only accept this report unreservedly. Although what happened on this occasion was of a scope without equal, it should be remembered that seismically generated changes in the level of the ocean floor have previously been known to far exceed anything recorded on land in either hemisphere. Logically, this would appear to be attributable to the fact that the earth's crust under the seas is far thinner than under the continents, hence more subject to deep fracture and extreme tectonic distortion.

Hewitt has this to say on the subject:

A curious and unexplained feature about the vertical movements caused by earthquakes is the excessive elevation sometimes recorded in the sea bed as compared with surface uplift. One of the largest occurred on a sea slope west of Greece, which was raised from five

[19]C. F. Richter, *Elementary Seismology* (San Francisco: W. H. Freeman and Company, 1958).

[hundred] to seven hundred metres in October, 1893. Such colossal submarine disturbances at least partly explain the great tsunami which cause such devastation.[20]

The above figures indicate a vertical displacement of 200 meters or 656 feet. Admittedly, this is no Sagami Nada, but it does seem clearly indicative, a significant measuring point on the scale of possibilities. [21] True, a few scientists, including more than one Japanese, do not accept the sea-floor displacement figures on the Kwanto tragedy. But suppose they *were* in error? Suppose a miscalculation *was* made? How large a mistake could it have been? The computations were not those of a single individual but of teams of experts working together, very experienced seismologists and oceanographers supplied with the most modern equipment available to the Japanese Navy at the time. Even if error existed, it could hardly have been profound.[22] And yet, with a submarine crustal fracturing as extensive as this undoubtedly was, indeed, as colossal as this, it was followed not by a civilization-destroying tsunami, but by a 30-foot wave whose ravages were restricted to the Yokohama waterfront. Why? The gross block movement here was mainly vertical, rather than horizontal, the throw of the fault far and away in excess of all known precedent. Conjecture is endless, but to date no hypothesis has proven definitive.

Japan, of course, continued to be seismically active. In 1930 the Idu earthquake struck, memorable chiefly for the eerie pyrotechnical displays which for once were universally observed. Three years later, the ocean bed moved again.

The Sanriku earthquake of March 3, 1933, was assigned a magnitude of 8.6 or 8.9, depending on whether the figures of the Stras-

---

[20]R. Hewitt, *From Earthquake, Fire and Flood* (New York: Charles Scribner's Sons, 1957).

[21]A "sea slope west of Greece" could only refer to the northwestern curve of the Mediterranean Ridge, the great rise which flanks the Aegean Sea on the south and bends around the west coast of Greece to penetrate the southern reaches of the Adriatic. The ridge is bounded by a trench on its north side, a fracture attaining a maximum depth off Turkey of 10,032 feet.

[22]Considering the great number of soundings taken, any error at all would appear out of the question.

bourg Seismological Institute or of Richter are accepted.[23] Like its terrible predecessor of 1896, it was an undersea shock. Similar, too, was its location and depth (100 miles out on the landward slope of the Japan Trench at 2,000 fathoms), and the fact that its slow, undulant nature left Japanese coastal dwellers largely unalarmed. The tsunami it generated ripped the Sanriku Coast from end to end, all 220 miles of it, attaining heights of 75 feet in some areas. Warnings were broadcast at the last moment, tragically late but far better than none. People remembered, bestirred themselves; evacuations were carried out. Still, 8,000 boats were smashed, almost 9,000 homes obliterated. When the death toll was tallied it was found that more than 3,000 people had died, every last one a victim of the seismic sea wave.

[23]Due to such variables as the distance between the earthquake focal point and the recording station, azimuthal anomalies, the lack of standardization of seismographs and seismographic conversion formulas, etc., differing magnitudinal values are often given for the same disturbance.

8

Tsunami of the Seventeenth and Eighteenth Centuries

Port Royal, Jamaica; Lisbon, Portugal

Probably because of its geographic nearness, to the American mind in particular very few places on earth are as romantic as the Caribbean. For the historically oriented the mere name conjures thoughts of gold and jewel-laden Spanish treasure fleets and the English and French buccaneers who preyed on them; of Ponce de León landing in Puerto Rico; Columbus in Santo Domingo; Velásquez in Cuba; Cortés in Mexico. Even the white man's discovery of the Pacific reminds us of this storied region, for it was from the Spanish Caribbean settlement of Darién that Balboa and his followers crossed the fever-ridden Panamanian isthmus to see, for the first time in 1513, *El Mar Grande del Sud,* that Great South Sea whose deceptive calm gave it the nickname *El Pacifico,* "The Peaceful."[1]

Even today the legendary romance persists. True, *hidalgos, conquistadores,* and pirates have long since commingled their dust with that of the centuries, while the scant metal remains of many a lordly galleon line the sea bottom, detectable to modern divers

---

[1] The colony founded by the Spaniards on the west coast of the Gulf of Urabá (the north or "Pearl Coast" of the Isthmus of Darién) was originally named Santa Maria La Antigua del Darién. In 1519, with the completion of the important Spanish trading center of Panama, it was abandoned.

only as sand-covered rows of cannon.[2] This makes no difference to the tourist. He travels the archipelago for the beauty of the islands, for candid glimpses of the carefree life of the people, for the excellent game fishing, the marvelous sun and invigorating ocean air. Calypso music? That, too. Or perhaps to find captivating bargains in the countless island bazaars. He correctly regards the Caribbean area as one of the foremost meccas in the world for the unhurried traveler, a paradise of waving palm trees and purple lagoons, of remote beaches tinted by sunsets rivaling those of the South Pacific.

But there is another way to look at the Caribbean; that is, to see it through the eyes of a geologist. To such an observer it is equally fascinating, equally challenging, but very different. For the geologist sees it without the superimposed trappings and architectural embellishments of man, without the cities, ports, and harbors, indeed without a trace that man was ever here at all. To the geologist, the region, viewed first from the global level, is circumscribed by one of the smallest of the planet's tectonic crustal divisions, the Caribbean Plate. He notes that in the eastern end of this sea its line of demarcation adheres closely to the island arc of the Lesser Antilles: viz., from south to north—Grenada, Saint Vincent, Barbados, Saint Lucia, Martinique, Dominica, Guadeloupe, Montserrat, Antigua, Saint Christopher, and Saint Martin. The northern edge of the plate cuts straight through the Greater Antilles, passing just north of the Virgin Islands, Puerto Rico, Santo Domingo and Haiti, then running southwest through the Windward Passage to divide Cuba from Jamaica.

To the south, the Caribbean Plate rather closely follows the curvature of the north coasts of Venezuela and Colombia, while its western edge, overriding Central America, abuts against the Cocos Plate along the line of the Middle America Trench. It is to be noted that the Caribbean Plate is nowhere subducted (forced beneath the crust), but is itself a segment of that enormous American Plate

---

[2]Contrary to well-established motion picture tradition, the intact hulls of such long-sunken ships are not to be found today by divers in tropical seas, though occasionally part of a keel or keelson or even pieces of frames are discovered. For the most part the wood is gone, salt water action and the teredo worm having caused it to disintegrate long ago. The remaining metal objects are usually so thickly encrusted with coral and other forms of sea life as often to be unrecognizable until cleaned.

whose northern and southern sections, continents in part, themselves subduct the Pacific Plate and its adjuncts along trench systems extending from central Mexico to Cape Horn.

At this juncture the untrained observer might easily make a mistake. To him, it might appear logical that the Caribbean Plate, thus seen to be "riding higher" than certain of its Pacific counterparts, would, seismically speaking, be relatively less unstable. Nothing could be further from the truth. While the eastern flank of this region does not reveal to echo sounders an especially deep-laid fissure, that which borders the north (actually, the northern edge of the plate itself) is ponderously and deeply faulted, as a study of the sea floor readily attests. Called the Puerto Rico Trench, the mighty rift runs all the way from a point about two hundred miles due east of Antigua to the southwestern corner of the Gulf of Honduras, the full length of the Caribbean, and immediately north of Puerto Rico (at 27,500 feet below sea level) contains the deepest spot yet found in the Atlantic.

It is *this* fracture zone that is seismically significant. In that part of it extending between Haiti and Cuba is a second deep of 26,417 feet, while its western half, gaping wider than most crustal faults anywhere, reveals a throw (vertical displacement) of dramatic proportions.

Geologically then, what sort of picture does the Caribbean present? In its eastern and south central regions are found the Venezuelan and Colombian Basins, running to maximum known depths of 16,594 and 12,926 feet, respectively. In its northwestern quadrant is the Yucatán Basin, a smaller but topographically similar depression joined to the Gulf of Mexico by the Yucatán Channel. Island arcs in the north and east; oceanic basins enclosed by them; running through the whole, knifing down to great depths within the crust, a gigantic fracture of largely unknown seismic potential. Historically, what has been the result? In the 13 years from 1947 to 1959, 40 earthquakes rocked the region, each of a magnitude higher than 6. Two were recorded at 7; two at 7.3. Four of them produced heavy casualties, and in areas of densely concentrated population the loss of life would have been catastrophic.

Were tsunami generated here? Four spring immediately to mind, waves that struck in the years 1842, 1907, 1918, and 1946. But there was another one much earlier. Because *this* wave and its preceding earth shock seemed to mark the end of an era, because for some that era was steeped in its own very singular brand of swashbuckling romance, historians and geologists alike tend to remember it. So, in their legends and folktales, do the hardy, laughing-eyed inhabitants of Jamaica.

It was the seventh of June, 1692, shortly before high noon. It was hot. Under a blazing Caribbean sun the broad reaches of Port Royal Bay shone like burnished copper, a bay more than half-enclosed by the long, westward-curving sandspit of the Palisadas Peninsula. Along the final stretch of the peninsula lay the fabled city of Port Royal, sometimes described as the "sole" of that odd-shaped formation, as the mangrove-choked tongue of land known as Gallows Point was the "heel." And Port Royal was infamous in this year of grace. Her denizens, mainly English and chiefly seafaring, numbered among them many "retired" freebooters, cutlass and pistol-bearing sea-robbers turned merchants, tavernkeepers, and *honest* seamen to appease the British Crown. The great buccaneer, Henry Morgan, was only four years in his grave, buried here; despite his betrayal of his own kind, a few remained who would honor his memory and quaff a noggin of rum to him. And they were proud, these English. Gazing across the bay at night, they could see the lights of Spanish Town. Always it reminded them of victory. When they'd wrested Jamaica from the Spaniards, that settlement was called St. Iago de la Vega. Now it was an *English* capital city, and the Blue Mountains which frowned down on it from across the bay were English, too.

The clocks of the town described as "the world's wickedest" ticked on. It was at night that the port really came alive, when its grog shops, taverns, gaming rooms, and brothels did a rollicking business, and weather-seamed, tattooed seamen swaggered about the narrow streets with their rolling "sea gait," the rum in their bellies making them "more than ordinary quarrlesome" and their hands ready at their sword hilts. Just now the place was quiet, very quiet. In the harbor, more than twenty vessels with furled canvas rode silently at their moorings, the sun-dappled water so still that even the creaking of ships' cordage could not be heard. The populace sweated. Some of the citizens dozed.

It was 20 minutes before noon. Along Fisher's Row on the town's west shore, a group of black freedmen lazed in the stifling heat near the turtle pens, seeking what shade they could find, while the shops in York and Tower Streets seemed nearly deserted. Down on the waterfront the forge of the King's Lane smithy was silent, the smith seeking repose. In the careening area near Watermen's Wharf the Royal frigate *Swan* lay hove down, canted over on her beam-ends to allow her fouled hull to be scraped and recalked. Nothing seemed to move anywhere. At Sir Thomas Lynch's wharf a gnarled wherryman argued halfheartedly with a king's customs officer. At Edward Watkin's west-side hostelry, the Sign of the Blew Bell,[3] a trio of British seamen eyed the serving wench speculatively, ordering double grog as they appraised her.

It was quiet at Saint Paul's Church, too. The chiefest of the town's several houses of worship, the imposing structure was served by its rector, the Reverend Emanual Heath. Mr. Heath had a visitor, John White, Acting Governor of the Island. As he accepted wine from his host, the Governor's beringed hand emerged from a spray of lace. It was good wine, the best wormwood the island produced, the peer of any they boasted on Barbados. Lifting his glass, the Reverend Heath appraised its rich color.

Abruptly, the earth convulsed. Beneath them the carpeted floor of the large, handsomely appointed vestry undulated as the roar of the ground shock smashed against their eardrums.

"Lord, sir, what is this?" cried the churchman.

Mr. White was trying to keep his balance. "It is an earthquake!" he shouted. "Be not afraid, it will soon be over!"

The above-quoted words are from a letter written by the Reverend Heath shortly after the catastrophe. He also wrote: "In the space of three minutes . . . *Port-Royal*, the fairest town of all the English plantations, the best emporium and mart of this part of the world, exceeding in its riches, plentiful of all good things, was shaken and shattered to pieces, sunk into and covered, for the greater part by the sea. . . ."[4]

Miraculously, both Heath and White survived, but another contemporary account gives an idea of the extent of the catastrophe.

---

[3]The odd spelling was characteristic of prevailing seventeenth-century illiteracy.

[4]From a quote in the *National Geographic Magazine,* vol. 117, no. 2 (February 1960).

In part, it reads: "The earth heaved and swelled like the rolling billows, and in many places the earth crack'd, open'd and shut, with a motion quick and fast. . . . In some of these people were swallowed up, in others they were caught by the middle, and pressed to death . . . the whole was attended with . . . the noise of falling mountains at a distance, while the sky . . . was turned dull and reddish, like an glowing oven."[5]

The first great shock was quickly followed by two more. Huge cracks yawned. In a few short moments of incredible violence the entire waterfront area slid toward the sea and was engulfed, carrying hundreds to their death by drowning. Built on the town's northwest promontory, Fort James crumbled and went under, her massive brick walls no match for the fury of the earthquake, while on the other flank Fort Carlisle did the same. Between them, street after street buckled, collapsed, and slid toward the sea: most of Fisher's Row with its turtle crawls and meat market; most of Lime Street; the full length of Thames, Queen's, and High Streets. The King's House and the King's Yard were between Thames and High Streets. They were crushed and disappeared. South of High Street on the east side of town was the solid edifice of Saint Paul's Church, with some of Port Royal's most noted landmarks stretching away on two sides of it. The bell tower of the church fell first. Then the spine of the structure itself broke and the building gaped like one of the flayed carcasses displayed on Fisher's Row. As the fractured and shaking land subsided, sea water rose to invade the ruins, inundating wharves, warehouses, waterfront taverns, Governor's House itself. But the worst was yet to come.

History fails to record who saw the tsunami first. Judging from the old accounts, Thames Street (also called King's Street), with its numerous business establishments, was already gone by the time the wave rose out of the sea. It came not from the open ocean but from Port Royal Bay itself, moving against the north shore of the peninsula, the most heavily populated, perhaps witnessed by a few hardy souls from various higher vantage points. No record of its size exists. From several known facts, however, we may draw certain inferences as to its power.

It seems to have destroyed most of Port Royal's shipping. One

[5]Ibid.

contemporary memoir speaks of "near total destruction of ye merchantmen and pinnaces." Another, written in the year of the catastrophe, stated: "Several ships and sloops were over-set and lost in the harbour. Amongst the rest the Swan Frigot . . . was forced over the tops of many houses . . . she did not over-set but helped some hundreds, in saving their lives."[6]

So far as the tsunami's strength is concerned, this "voyage" of the *Swan* tells its own story. A typical Royal Navy frigate of the day was a three-masted, square-rigged vessel with raised poop, quarterdeck, and forecastle, a craft displacing anywhere from 8 to 12 tons. Port Royal's many buildings were structures of two and four stories. The *Swan* was picked up by the wave and carried completely over them. How she managed to save "some hundreds" is not entirely clear. Before heaving her down on the careening ground they would have unshipped her guns and possibly lightened her ballast as well, so altering her trim as to make her unmanageable even in a normal sea. Yet accounts of her experience, though sparse concerning details, are well-authenticated. Logic would indicate that the tsunami "outran" her as it roared over the town, initially lifting and thrusting her forward, then passing underneath and beyond her in a way that never would have happened if the frigate had been firmly in the wave's grip. After that it was a matter of luck, quick thinking, and superb seamanship.[7]

Two-thirds of Port Royal was destroyed. Two thousand people died. Due to ground movement or the savagery of the wave, the cemetery where Henry Morgan and others were buried behind the town was obliterated, the sea at last claiming the remains of the victor of Panama and Maracaibo. One might speak of fascinating events, of Americans like Edwin Link and Robert Marx who in recent years dove on this storied site, a city unaffected by man or the elements, unchanged over the centuries, sealed in a moment of time. One might mention the artifacts recovered, the brass watch, for example, which, after X rays were taken of the impressions left on coral encrustation by its missing hands, revealed to scientists

---

[6]Ibid.

[7]Even though she was heeled over at the time the wave struck, clearly some seamen must have managed to get aboard her, perhaps through inadvertent entanglement in her standing or running rigging.

the exact time of the quake: 11:43;[8] the cannon and cannon balls, ballast bars, cooking utensils, pewter plate, bricks, roof tiles, clay churchwarden pipes, onion bottles, sword hilts, knife handles, candlesticks, axe heads, and various other tools, and the ship's yard.[9] All of these relics are very interesting and informative. They and others may be seen today in the Institute of Jamaica at Kingston.

But the traveler should not be deceived. When he visits Port Royal as it is now, a thoughtful analysis of what he views brings realization of total change. The old city he cannot see unless he dives on it and searches the murky waters of the harbor, nor much of it even then. He may walk the streets of its successor, a sleepy little town where nothing ever happens; he may explore; but, inevitably, the finality of the earth shock of 1692 will make itself felt, giving emphasis to the laconic announcement on Herman Moll's Admiralty chart drawn in 1720: "Port Royall Sunk."

The historically minded person may walk the road east of still-existing Fort Charles, a way paralleling the series of earthworks once called Morgan's Line, pondering on the grave no longer there, on that wild breed of men who held this redoubt in the midst of Spain's hostile American Empire, defying the Dons and hell itself, laughing in their defiance. He may think of Horatio Nelson, once commandant of Fort Charles, a young man then unknown, whose name was later to be emblazoned in the annals of naval history as no other.[10]

But the geologist will think of something else. Again he will turn to the map of the Atlantic Ocean floor, his fingertips moving to the Caribbean, to that vast fault splitting it from end to end, the broad,

[8]This remarkable find was credited to Al Barnesky, one of the divers sent by the U.S. Navy to assist the Link expedition.

[9]This spar, 20 feet long and bound by not a single fitting, had been preserved from the action of sea water and various marine borers by a coating of fine clay. Together with a fifteenth-century swivel gun and breechblock, it was found in 30 feet of water at the point where King's Lane intersected Thames Street.

[10]Nelson commanded there in 1779, when the sea washed the fortification whose approaches are now silted in. Other Caribbean islands associated with the famous admiral's name are Nevis, in whose Fig Tree Church he married Frances Nesbit, and Antigua, where at English Harbor (in the facility since named Nelson's Dockyard) he refitted the British fleet for the voyage back to Europe and the decisive naval battle of the Napoleonic Wars—Trafalgar.

very deep, and largely unexplored Puerto Rico Trench. Between Cuba and Jamaica it lies, between Cuba and the single island forming Haiti and the Dominican Republic, passing far beyond them to the east and west. What does it portend for the future? The geologist cannot know, can only speculate, aided by existing knowledge. As he looks at it, the lines of his face may set somewhat grimly. Catastrophic earthquakes in some as yet unknown age? Without any doubt. Bigger, more destructive tsunami? Equally a likelihood. But man has always found a way of dealing with his environment, of coming to terms with it and surviving. Natural disasters will recur here as elsewhere. But the golden beaches will remain, the whispering palms, the quiet lagoons, the incomparable sunsets. And somewhere, for some men, there will always be a lovely island maiden, waiting—with a hibiscus in her hair.

As the reader will comprehend, the above-discussed Caribbean is a part of the western Atlantic. When it comes to the eastern Atlantic, say, roughly from 35° to 45° north latitude, the picture is very different. Here, off the west coasts of Spain and Portugal, there is no great trench system. Instead, between the east-west thrusting Azores–Gibraltar Ridge to the south and the Biscay Abyssal Plain off France, a large depression dotted with many seamounts extends seaward to the series of latitudinal faults which so thoroughly fracture the stupendous east slope of the Mid-Atlantic Ridge. This depression is called the Iberian Basin. To date, its maximum recorded depth is 17,700 feet. One of its elevations, the Ampere Seamount, is so high that its summit, rising from a depth of 15,868 feet, comes to within 132 feet of the ocean's surface.

Examining the Iberian Basin, the geologist is struck by the difficulty of accounting for the great catastrophe that struck the City of Lisbon, Portugal, in the middle of the 18th century. True, a mighty rift valley exists in the central Atlantic and is heavily fissured from east to west, but this seismically unstable area is hundreds of miles out from the west coast of the Spanish peninsula. Having intensely studied the basin itself, the scientist begins to trace its edges. Then he notices something. The Azores–Gibraltar Ridge, although plotted on most charts as extending only from the Strait of Gibraltar to the Azores Islands, is actually much longer than that.

More accurately, it appears to be a part of something else, a gigantic fault system that begins in the central Mediterranean near the toe of the "boot of Italy," passes through the Strait of Gibraltar, and runs all the way to the Mid-Atlantic Ridge.

Now it becomes more apparent. If one compares a chart of the Atlantic Ocean floor with a map showing the location and extent of tectonic plates in the eastern hemisphere, the Azores-Gibraltar Ridge stands out clearly for what it is, an important segment of the northwestern edge of that colossal crustal division called the African Plate. Does that throw light on the situation? Studies indicate that all along its northern edge, from the Coast of Syria to the Mid-Atlantic Ridge, the African Plate is being subducted by the Eurasian Plate, a process which ages ago produced the zone of vulcanism containing such still active vents as Thera (or Santorini), Vesuvius, Etna, and Stromboli. Is the degree of vulcanicity in this region matched by a similar seismic potential springing from the same crustal anomaly, a seismicity extending well into the Atlantic? The answer would seem to be yes. Although there were no seismographs when it occurred, the Lisbon earthquake, judging from its effects, was the most violent in history, a shock to which both Gutenberg and Richter have tentatively assigned the unheard-of magnitude of 8.75 to 9.[11] It was felt over a million square miles. It appears to have had at least two foci, one of which, in North Africa, caused ground motion that heavily damaged Fez and Mequinez, Moroccan cities more than 400 miles from Lisbon. Although other quakes, particularly those in China, had taken or were to take a greater toll in lives, so far as is known no man in historic times had seen its equal for sheer power. It is this extraordinary phenomenon we shall now examine.

---

[11]Beginning on page 96 of his excellent study, *When the Earth Trembles*, (New York: Harcourt, Brace & World, Inc., 1964), Haroun Tazieff writes as follows: "I must dwell once more upon the prodigious energy of a magnitude 8 earthquake—energy in the nature of 10^{26} ergs, a million million kilowatt hours, released in a few seconds—and upon the comparative rarity of these shocks; for out of the fifty or sixty million earthquakes that occurred in the world between 1904 and 1960, scarcely sixty reached this awful figure."

*   *   *

It was the autumn of the year 1755, November 1, All Saints' Day. In the thriving port of Lisbon much of the population had crowded into the cathedrals and churches to celebrate morning mass, having first dutifully watched the religious processions in the narrow streets and heard the mournful chanting of the penitential *saetas* (chants dealing with the Crucifixion). Throughout Portugal the priesthood was still in firm control, as it had been for centuries, and especially here in the capital city. Friars and monks of the various orders were to be seen everywhere, as were religious mendicants. Most of the larger buildings were monasteries, convents, or churches, and the huge bronze bells in the campaniles clanged continuously.

Lisbon was a city built on hills, seven of them. Higher up on the slopes lived many of the poorer people. As was generally the case with their class throughout Europe, their lot was harsh, their privation compounded by extreme congestion and squalor. When, on days the Church allowed them to work at their livelihood, they took a moment from their labors, they gazed down on an impressive panorama. At the lower levels the buildings were much larger; near the center of the city's waterfront the land was virtually flat and only a few feet above sea level. Here rose the Royal Palace, an imposing edifice of pink marble. Ostensibly, it was presided over by the weak King Joseph. Actually, like the country itself, it was ruled by a remarkable man who held the offices for both foreign affairs and war, Sebastião de Carvalho, Marquês de Pombal, the would-be master of Portugal.

Looking out across the broad reaches of the Bay of Oeirás, actually the estuary of the Tagus River, seeing the large ships of many nations there, the prosperous plying of trade in lighters and small sailing craft, the stranger might have been unaware of the political ferment ashore, the deadly intrigues in high places, the vicious and continuing struggle for power. Letting his eyes wander from the palace to the Tower of Belém on its rocky point, he would think of Portugal's greatness, of Dias sailing from here to round the Cape of Good Hope, Da Gama to reach India, Cabral to discover Brazil. Yet, within the teeming city, Pombal, hated by the nobles and the priesthood for his reforms, pitted toe-to-toe against the savage oppression of the Holy Inquisition, waged a perpetual war for existence.

It was 9:35 A.M., the day fair and the temperature brisk. Visiting foreigners might have remarked other evidence of this foremost of all holy days: the fresh paint on signboards, doors, and lintels; the scarlet-hung balconies; the sand spread in the squares and all the larger streets; the flag-bedecked windows; even the occasional splotches of blood where zealous penitents had had themselves pierced with sharp instruments or flogged with barbed whips. The entire city seemed alive with sound. The massed voices of the great cathedral choirs rose above the sonorous ringing of the church bells. People gloried in it or were held in awe, some crossing themselves before the act of merely stepping outdoors, others falling on their knees in the streets to pray. For once, even the city's fencing masters, much exalted by the militant Pombal as in the old days, could be found in attitudes of humility. Their proud heads were bowed in church.

The minutes passed. Nine-thirty-six came and went. Nine-thirty-seven. While, between the periodic chanting of high-pitched *castrato* voices, clergy and congregations intoned their ancient litanies, deep underneath them was movement. They weren't aware of it as yet. The widespread dilatancy of underground rock is a subtle thing. Yet somewhere along the north edge of the African Plate the infinitesimally slow process of subduction had quickened; a major segment of the crust was slipping. Nine-thirty-eight. In lightning-swift chain reactions the motion spread laterally to adjacent regions, causing rock to swell and crack beneath the west coast of Iberia and the north coast of Morocco, generating sympathetic crustal vibrations far out to sea. The pressure built and built, trillions of tons of it, the stored energy doubling and trebling with the passing seconds, reaching incalculable levels. Nine-thirty-nine. Rock liquefaction was beginning, starting with the softer clays, moving through sandstone and limestone into the harder crystalline minerals. The hands of Lisbon's clocks stood at 9:40.

Without warning, the first mighty ground shock struck the capital. While the ships and boats in the bay were jolted as by a mammoth sledge, their startled crews looked shoreward to see an appalling sight. In a nightmarish phantasmagoria the city of 250,000 souls seemed to be dissolving. First to fall were the tall towers and massive stone belfries of the cathedrals. They smashed through roofs, tons of hurtling granite that crushed the people inside, the thunder of the impact adding to the continuous earsplitting roar of sundering rock. In the universal din no screams could be heard. A tremendous dust cloud rose above the scene like something out of Dante's *Inferno,* blotting out the sun, leaving the frantic survivors to choke in the gathering darkness. The shaking slowed, its violence nonetheless increasing while the incredible vertical movement shifted to an even more hideous lateral thrust. The compressional effect of the primary wave train's undulant motion had shattered buildings to the roots of their foundations. Now the secondary waves, moving transversely and wrenching the structures at right angles to their course, brought them crashing down.

Among the battered survivors still able to extricate themselves a pitiful pandemonium ensued. The maimed, the half-crushed, and the blinded shrieked in their agony. Mothers dying from their own injuries snatched at the bleeding bodies of babes and small children, gasping for air in a world of smothering dust, struggling toward wisps of imagined light while the broken ground beneath them rolled and heaved, and yawning fissures opened.

Along the waterfront every building was down, the royal palace a dust-shrouded pile of twisted wreckage, its flanking edifice crumbled as though hit by titanic explosions. Fires were starting. Throughout the ruined city thousands of devotional candles on scores of altars had set church tapestries and altar-cloths alight, while overturned hearths in homes had tumbled glowing embers onto floors of wood. The fires spread rapidly through the vast heaps of rubble. There was no available water to fight them, no citizens' organization equipped to do so in any case. People fended for themselves, panic breaking out in widely separated areas, blind terror governing many, tens of dozens trampled to death wherever wild-eyed throngs of frantic survivors could find sufficient space in which to run.

Ten minutes after it had begun, the shaking stopped. People now filled the streets and squares, the adrenalin summoned by insensate dread infusing them with an incredible strength as they made their way in thousands toward the waterfront, seemingly impelled by a universal instinct. Over the broken remains of buildings they climbed, oblivious to tottering walls and still falling masonry, ignoring the rising flames. To escape by sea was the all-pervading purpose (as it had been 17 centuries earlier at Pompeii and Herculaneum), somehow to reach the water, find a boat, get out to the waiting ships. Miraculously, several hundred made it, or so they thought. Finding everywhere along the bay front only an impenetrable shambles, they crowded onto the marble quay, doubtlessly taking comfort and assurance in its ponderous strength.

For a short while they found refuge along this solid structure, though many continued in great agitation and not a few were toppled into the water. Then the earth began to heave again; the characteristic roaring started; deadly crevasses opened as further splitting of the crust continued. The people now lost heart. Women wailed. Many persons of both sexes fell to their knees and prayed. Abruptly, beneath the quay's foundation the loose fill flowed out and the rugged wharf collapsed, casting its human burden into the sea. The rolling harbor waters closed over them, stifling their shouts and screams. None of them were ever seen again.

But nature's grim joke had not yet been played out. As the fire spread throughout Lisbon, rapidly becoming a holocaust, it first created its own vacuum, then, to fill that unnatural void, a fiercely rising wind. People fled in a frenzy before it. Finding themselves trapped on three sides, cut off by a roaring, red hot furnace, they sought the final alternative, struggling in desperation to reach the already crowded waterfront. This all took time. According to some eyewitness accounts, about 40 minutes passed before the waters of the harbor receded. Others claimed it was closer to an hour. In any case, the thousands who crawled like insects over that broken and debris-heaped shoreline were shortly exposed to a strange and fearful sight. Suddenly and with fantastic speed Oeirás Bay all but emptied itself, causing craft of varying tonnages to bottom and heel over, stranding hundreds of wriggling fish, exposing rocks never before seen by man to a distance of more than a mile out. It was the warning. In that vast and hapless throng of wretched humanity very few comprehended its meaning. They watched, fascinated.

Only minutes passed before the wave was first seen. Reliable surviving chroniclers computed its height at more than 50 feet. The masses panicked, stampeded. In thousands they tried to retreat into the city's flaming ruins, preferring to roast rather than face this primal horror. The rubble delayed them. The unbearable heat drove them back. Exhausted, resigned to a nameless fate, they turned to face the racing tsunami, seeing its black waters rise hellishly now as the rapidly shoaling bottom piled its awful crest higher and higher. Then it broke over Lisbon's lower reaches. More lofty than a five-story building, its thunderous mass carried all before it, hurling great ships against the already shattered structures, leveling them, drowning and crushing additional helpless thousands. It swept more than half a mile inland, an irresistible, raging surge bringing overwhelming death. Large parts of the city were underwater. Minutes later, many who had somehow survived the initial onslaught were caught in the all-embracing, murderous sweep of the backwash and drowned. The toll of lost lives was enormous.

Thus was much of Lisbon destroyed. Two more waves struck about an hour apart, finding little left to ravage. Close to high noon came the first violent aftershocks, but in a city already so stricken they did little additional damage. Fifty-three structures belonging to the palace complex joined that royal edifice in oblivion. Thirty-two churches perished, among them the greatest of the capital's cathedrals. Out of the city's 20,000 houses a bare 15 percent survived, heavily damaged, and these mainly on higher ground. The cost in flesh? In a day when any population census was itself invariably untrustworthy, casualty figures differed markedly. Estimates ran all the way from 30,000 to 70,000 dead, an unknown number of injured. The consensus of modern historians is close to 60,000.

Both Pombal and the king escaped. When the quake struck, King Joseph and most of his courtiers were riding in carriages between Lisbon and the multi-turreted Tower of Belém, the monarch having decided to take up temporary residence in the latter. Fortunately, Pombal had been exercising his horse on the city's outskirts. Now he pounded after his sovereign, finding him much distraught by events, close to tears, as irresolute as ever. The king sought his advice. Pombal warned that the devastated capital was no place for him now, and strongly urged him to take refuge in the tower. When the shaken ruler hesitated, asking what might be done for the people, Pombal made a famous reply.

"Sire," he said firmly, "we must bury the dead and feed the living."

Thereupon, Joseph granted his chief minister absolute power. Pombal returned to the city, used the army to take control, and quickly reestablished order. Looters he executed out of hand, many of them. Then, calling on the prodigious energy for which he was renowned, for weeks and months he labored like a man possessed, clearing away the shambles of his beloved city, consulting architects and builders, importing at heavy expense the raw materials to construct a larger and greater Lisbon, a metropolis *resistant to earthquake*. The clergy continued their war against him, their latest grievance the mass removal of thousands of corpses for hasty burial at sea without the final rites of individual sanctification. Pombal ignored them. His resolute act had almost certainly prevented the outbreak of typhus and other deadly diseases. It was enough for him.

This is not an historical treatise. Actually, there would be little justification for throwing light on Pombal herein, great man though he was, were it not for the important role he played in the history of seismology. For the first time, thanks to him, an earthquake was intensively studied in Europe. In France the following year, though Voltaire might publish his "Poem on the Disaster of Lisbon," subsequently enhancing his reputation with the satirical novel *Candide,* which briefly deals with it, it was Pombal, not Voltaire, whose indefatigable efforts and dedicated single-mindedness instituted in the Western Hemisphere the lifesaving practice of seismic investigation and analysis. His studies of the nature and extent of the damage, the field trips he personally undertook to examine and plot fissures, his carefully kept observational notes, his consultations with Japanese scholars sent by the Tokugawa Shogun to aid him— all these efforts and more he pursued for the sake of his people and that of mankind.

His power lasted 22 years longer, until the death of Joseph in 1777. Then a new sovereign ascended the Portuguese throne, Maria I. She was under the control of the clergy and nobles. She feared Pombal's influence, despised him for his ultimate curbing of the power of the Inquisition.[12] Before the year was out he was arrested on loosely contrived charges. His enemies demanded his death, but

---

[12] By subordinating it to the royal authority, Pombal dealt a crippling blow to the Holy Office in Portugal. When he expelled the nearly omnipotent Society of Jesus (the Jesuits) from both the homeland and the colonies, thereafter redrafting the property laws to reduce the wealth of the Church, he virtually eliminated clerical suzerainty.

to that level the Queen refused to lower herself. Pombal was stripped of his offices and exiled to his country estate. There, carefully guarded, a lonely man in an all-but-empty castle, he lived five more years, dying in 1782 at the age of 83.

Already mentioned was the fact that the shock which struck Lisbon also ravaged parts of North Africa. According to one report from Morocco, in an area somewhere south of Fez 10,000 people were engulfed by the opening of a gigantic fissure. Since this would be barren desert country drained by the rivers of the Grand Atlas, and since, obviously, no major population center existed between Fez and Rabat, in the same latitude, and Marrakech to the southwest, this claim may be discounted. Nonetheless, the reports of great tsunami striking the Moroccan Atlantic coast are well-authenticated, as are the casualty figures totaling between 9,000 and 11,000 dead along that extensive littoral.

With regard to seismic sea waves, the Madeira Islands should be mentioned. Of the two inhabited ones, Porto Santo seems to have escaped largely unharmed, while Madeira was very heavily damaged by a 50-foot wave. In its capital city of Funchal, waterfront structures were demolished and the old accounts speak of hundreds of fish left stranded in the cobbled streets. A glance at the map will show that this island group's location is less than 200 miles south of the Azores-Gibraltar Ridge.

Intriguing because of the complete lack of any documented details in extra-Hispanic literature is the report that this great shock sent a 60-foot wave against the Spanish Andalusian seaport of Cádiz, of prime importance to Spain since even before the Armada. Had it involved instead the destruction of a small town or fishing village the paucity of information would not be unusual. But Cádiz? Seldom has Spain's insularity been more evident. The magnitude of the devastation a 60-foot tsunami would wreak against that city's extensive waterfront and crowded shipping can only be imagined.[13]

---

[13]Even Great Britain, over a thousand miles from Lisbon, was hit by waves of from 9 to 12 feet in height. Large ships in harbors parted their mooring lines. Many small craft were seriously damaged and some were sunk. The opening of at least one large fissure on English soil was positively verified by ⁺he authorities.

By now the reader is familiar with the long reach of seismic sea waves. Therefore, it will come as no surprise that the same phenomenon which inundated Lisbon raced across the Atlantic to enter the Caribbean and make itself felt in the West Indies. The island of Saba lies 3,545 miles from the Iberian Coast. Yet dispatches from this extinct volcano which the Dutch colonized in the Leewards spoke of waves between 21 and 25 feet in height. They had no effect at Saba so far as habitations were concerned. Its coasts are sheer cliffs rising to 2,820 feet. Its capital, distinctively named Bottom, was built at 800 feet above sea level on the floor of the crater.

Of all the effects of the Lisbon earthquake, among the most interesting were the *seiches*[14] (oscillations in distant coastal and inland water bodies) which it generated in widely separated areas. Because they have at times been confused with other waves of seismic origin, the next chapter will deal with them and with various related phenomena.

---

[14] The word is pronounced sāsh in both singular and plural.

9

Other Forms of Wave and Some Clear-Cut Distinctions

Seiches, bores, and whirlpools. "Seismic waves" as distinguished from "seismic sea waves." The "ultimate wave"—Lituya Bay, July 9, 1958. The phenomenon of swashes.

Because of a multiplicity of factors, many of which operate in seeming contradiction to one another, any in-depth study of worldwide tide systems is certain to be a complicated undertaking. In this section, therefore, we'll limit our examination to four effects, as follows: natural tidal oscillations within certain water bodies; nontidal oscillations, or seiches, which may affect the former through extraneous kinetic influences; a word concerning tidal bores; oceanic and fresh water whirlpools.

When, on a windless day, we gaze on the unruffled waters of a lake or breakwater-enclosed harbor, no apparent motion of those waters may be discernible, least of all that to-and-fro action termed *oscillation.* Yet, in the quietest of such fluid bodies, oscillation *is* present at all times. This movement is due to a combination of causes: to the inertial forces affected by the earth's rotation and the gravitational attraction of the sun and moon. This movement, of course, is measurable. In the most literal sense of the word, then, we observe that even in the most insular and landlocked of water-filled basins a tide exists.

Not all bodies of water oscillate at the same rate of speed, a factor as true for completely enclosed basins as it is for those open to outside influence. The greater a basin's length, the longer its period

129

of free oscillation; the more shallow its depth, the shorter its period.[1] From this, mathematical formulas have been devised for calculating the natural oscillatory period for any given basin, establishing tables of figures which can then be compared with visual observations.[2] If the computations are done correctly, the comparisons are usually very close.

Basins, of course, take many different shapes, some entirely closed to the sea and to all other water bodies, others open at one end or at both. Where there exists a lake or inland sea, closed completely except for rivers or streams which may feed it, any waves that form will tend to oscillate round a stationary axis in the center, a point defined as the node. At this point, the horizontal flow at mid-tide reaches its maximum while the water level remains constant. At the opposing ends, on the other hand, no horizontal flow occurs, but the rise and fall of the water attains its maximum height.

In the second type of basin, open at one end, the node is created at the opening. Here the current will flow at maximum strength, while at the enclosed end an antinode is created.

In the third type, open at both ends, two nodes develop at opposite extremes.

Thus far, all references have been to natural tidal oscillations. When the motion is *induced* from a source outside the water body in question, it may come from a rising wind, changes in barometric pressure, or the vibrations imparted to crustal rock by an earthquake. Such oscillations are nontidal and are generally termed *seiches.*

According to Webster, *seiche,* a word derived from the French-speaking cantons of Switzerland, is defined as "an oscillation of the surface of a lake or landlocked sea, varying in period from a few minutes to several hours."[3] To the degree the definition is incom-

---

[1] *Free* or *tidal* oscillation refers to those movements within a water body which are present at all times. *Forced* or *nontidal* oscillation is movement generated in the same body by some form of outside disturbance.

[2] The length and depth of the basin are the sole factors in this calculation. The breadth is irrelevant. For example, Long Island Sound is 80 miles long and has an average depth of 10.8 fathoms. From these figures it can be determined that its natural period of vibration is 42,550 seconds or 11 hours and 49 minutes. Similarly, that of Lake Geneva is 4,464 seconds or 1 hour and 14.4 minutes.

[3] Seiches first came under scientific scrutiny in 1870 when the Swiss physician, F. A. Forel, was investigating Lake Geneva's stationary waves.

plete it is inadequate, for in addition to "landlocked" water bodies seiche action may affect the fluid dynamics of harbors, bays, headland-protected coves, and breaking inlets. Even the water of rivers and canals is not at all immune to seiche influence.

Although the generation of very large seiches is a comparatively unusual occurrence, throughout history they have been known to occasionally accompany severe seismic disturbance. This applies only to the great tectonic shocks, shallow of focus and caused by crustal slippage. In the case of volcanic earthquakes, their eruptive foci are so deep (down to 45 miles) and their vibrational periods so short that even the most violent of them do not produce seiches except, perhaps, in the rarest of instances.[4]

As will be recalled from foregoing chapters, earthquakes generate wave trains, P waves, S waves, L waves, etc., consisting of hundreds of very intense, far-reaching vibrations, each one oscillating at a given frequency. When the frequency of any such train happens to coincide with the frequency of natural oscillation of a particular body of water, it reinforces the effect through what is called *sympathetic vibration* or *concomitant resonance,* imparting a rhythmic motion to the entire fluid mass. In small basins the effect is small. This is because the period of natural oscillation is too brief for coincidence and rhythmic reinforcement. In large basins, however, where induced frequencies sometimes match and augment previously existing ones, the establishment of resonance produces much higher tides.

In the majority of instances, earthquake-induced seiches do not occur close to the epicenter of a great shock but hundreds of miles away. This is due to the following factors: It is in the nature of the transmittal of earthquake shock waves that only those close to the epicenter consist of high-frequency vibrations, while those at much greater distances are of lower frequency. High-frequency waves

---

[4]An example of the failure of titanic volcanism to generate seiches would be the five-month-long cataclysmic eruption of the volcano Bezimianny (meaning "without a name") on the Soviet Union's Kamchatka Peninsula, the mightiest volcanic outburst of this century. The eruption began on October 22, 1955. Prior to the final paroxysm on March 30, 1956, ground movement exceeded 1,000 microns (1 millimeter) at Klyuchi, 40 miles distant. On the final day, the mountain truncated itself in one of the greatest blasts of modern times, losing 590 feet in height, hurling millions of tons of ejecta to an altitude of almost 28 miles. Yet no seiches were reported.

often are of too short duration to induce seiche action. Low-frequency impulses, on the other hand, moving as slowly as 2 miles per second (as against about 5 miles per second for the others), have that lower velocity and higher persistence more likely to augment rhythmic movement in a body of water, sometimes to an amazing degree.

To cite specific examples of seiche action, we might begin with that which is unquestionably the best known to science, the Lisbon earthquake previously described. Initially, the enormously far-reaching scope of this shock should be recalled, a disturbance of unprecedented magnitude felt over an area of a million square miles, an effect so profound that in European cathedrals more than 1,000 miles from the Portuguese capital the massive crystal chandeliers were set in motion. However, it was primarily the seiches occurring in lakes, harbors and canals throughout western Europe that affected the reasoning of the English physicist, John Michell, geology professor at Cambridge. Ignoring the various notions then prevalent, by 1760 he had worked out the first scientifically based theory on the transmission of earthquake shock waves.

Probably due to the distance factor previously observed, no seiches were reported closer to Lisbon than 500 miles, nor any farther away than 1,500. However, within those limits every water body in western and northern Europe was set in motion, from Finland's 10,000 lakes to those of Switzerland, from the Scottish lochs and firths to the fjords of Norway, from the Zuider Zee of the Netherlands to the rivers, lakes, and canals of western Germany. In such teeming Dutch ports as Rotterdam and Amsterdam, large ships parted their mooring lines, as they did at both Dartmouth and Plymouth on the distant coast of Devonshire.

The long and violent history of seismic instability on this planet offers many examples of seiche action, some of them producing very singular effects. On December 16, 1811, a great earthquake centered near the town of New Madrid, Missouri Territory, took its name from that small community, but additionally affected the territories of Illinois and Arkansas, and the states of Tennessee and Kentucky. One authority has written about it as follows: "The rivers themselves were violently agitated by the earthquakes. Where bluffs formed the river banks, large landslides occurred, completely changing the navigable channel of the rivers. Several islands in the

Mississippi vanished completely; one is said to have taken a colony of river pirates to their death. Temporarily the Mississippi reversed its course. This may have been due to the rise of the land in the areas already described, or to the temporary blocking of the channel by landslides. The surface of the waters was thrown into violent waves which, rushing onto the land, did great damage to timber."[5]

Although no means of calculating exact magnitude existed in 1811, it will be noted that this shock, like that which destroyed Lisbon, was felt over a million square miles of territory, throughout the entire length of the Atlantic seaboard and from Canada to the Gulf of Mexico.

Many westerners in the United States still recall the earthquake that struck Hebgen Lake, Montana. The National Geographic Society's Samuel W. Matthews described it memorably:

Late that clear and gentle night a full moon rode above the tumbled mountainland where Montana, Wyoming, and Idaho meet. Along the mile-high Madison River the air hung cool and still. Monday, August 17, 1959, was nearly over.

Suddenly, 22 minutes and 45 seconds before midnight, an earthquake of historic force wrenched thousands of square miles of the northern Rockies. The earth's crust, warped by unimaginable strain, broke and shifted along several faults, or fractures. A lake in southwestern Montana tilted. Water leaped a dam in racing waves.

Downstream in the Madison's narrow canyon, a mountain lake 1,300 feet high and half a mile wide split off and came roaring down on sleeping campers. Eighty million tons of rock and earth smashed into the river and splashed 430 feet high against the facing slope. Water and terrible blasts of air shot out from under the thundering slide.

In Yellowstone National Park, twenty miles east, the peaks trembled, rockfalls cascaded across roads, and steaming geysers, mud volcanoes, and boiling-hot pools erupted simultaneously in unseen splendor.[6]

[5]John H. Hodgson, *Earthquakes and Earth Structure* (New Jersey: Prentice-Hall, Inc., 1964).

[6]S. W. Matthews, "The Night the Mountains Moved," *National Geographic*, vol. 117, no. 3 (March 1980).

Here again, the action of seiches is illuminating.

> Displacement on the fault scarps in all instances suggested the depression of Hebgen Lake and its environs. This regional depression was confirmed by lake soundings and by the rerunning of precise levels. The lake bottom was lowered by an average amount of about twenty feet. . . . An eyewitness standing in the moonlight on Hebgen Dam and looking down its sloping face could not see the surface of the water, so far had it receded. Then with a roar it returned, climbing up the face of the dam until it overflowed the top, and poured over it for a matter of minutes. Then the water receded again, to become invisible in the moonlit night. The fluctuation was repeated over and over, with a period of about seventeen minutes; only the first four oscillations poured water over the top of the dam, but appreciable motion was still noted after eleven hours.[7]

Clearly, in the case of the Hebgen Lake shock the factor of seiches being usually absent in close proximity to an earthquake did not apply, since the epicenter was within a few hundred yards of the lake's eastern shore, almost on the Wyoming border. Although not heretofore mentioned, it was equally inapplicable in the great earthquake in Chile in 1960.[8] In that catastrophe, seiches with amplitudes exceeding 3 feet were seen in some of the stricken nation's basins, particularly in the Chilean Lakes Region in the vicinity of the Reloncavi Fault. The reader will recall that in this disaster there were many epicenters, their reach extending over nearly 12 degrees of latitude. The closest ones to observed seiches, however, were less than 50 miles away.[9]

As has been said, seiches are *induced,* and, in a quite literal sense, nontidal. Were this not so, the term would have some synonymity with *bore* but it does not. In an elementary but nonetheless correct fashion, Webster defines the word *bore* as follows: "A tidal

---

[7]Hodgson, op. cit.

[8]See Chapter 5.

[9]The catastrophic Assam Earthquake (amplitude 8.6) of August 15, 1950, on the other hand, confirmed the general rule. Felt over 1.1 million square miles (in India, Pakistan, Tibet, China, and Burma), it generated seiches within roughly the range limitations of the Lisbon shock.

flood with high abrupt front, due to a rapidly narrowing inlet or channel, as at the mouth of the Amazon."[10]

Since, unlike seiches, they are tidal, bores recur in given areas with timeless regularity, and over the centuries a few of the more spectacular ones have become world famous. In eastern China, for example, there is the bore of the Fuch'un (formerly, the Tsientang Kiang), a river that empties into Hangchou Bay on the East China Sea. At the river mouth where the channel is broad, the vernal tide increases the depth to about 12 feet, but as the bore advances into the narrows its speed reaches a formidable 13 knots and the crest rises until it becomes a vertical wall of water 25 to 30 feet in height. At this point it is moving over 2 million tons of water a minute *against* the river's current, a force of truly awesome proportions. It is extremely dangerous. Over eons of time it has taken many lives, and the fishing boats and commerce-carrying junks it has claimed are numberless.

Curiously, centuries ago Chinese river men found a way to make use of the Fuch'un bore. Keeping their boats out of its sweep by mooring them in revetments dug along the river banks, they wait for the bore to pass, then enter its powerful afterflow and allow their craft to be literally dragged upriver by the suction.

In 1888, the British admiral Osborne Moore, then a captain commanding her Majesty's survey ship *Rambler*, wrote of this practice as follows: ". . . no less than thirty junks swept up in the after rush and passed Haining with all sails set but with their bows pointing in every direction, several proceeding stern first at a rate of ten knots toward the city of Hang Chow."[11]

Let's look at the definition again: ". . . a rapidly narrowing inlet or channel." This necessary distinction for the existence of a bore is seen when a comparison is made between the English Channel and the Bay of Fundy in Nova Scotia. They have similar dimensions. The English Channel is 123 miles in length, with an average

---

[10]This bore does not actually invade the Amazon but the Río Tocantins which empties into the Bay of Marajó, a part of the Amazon's enormous estuary. Located within the estuary and nearly filling it is the Ilha de Marajó, an island larger than Belgium. The Tocantins bore, called the Pororoca, reaches a height of 15 feet and moves upstream for an amazing 200 miles.

[11]Quoted from F. G. Walton Smith, *The Seas in Motion* (New York: Thomas Y. Crowell Company, 1973).

depth of 25 fathoms; the Bay of Fundy is 175 miles long, its depth an average of 50 fathoms. Both have a period of free oscillation or resonance closely approaching that of their semidiurnal tidal periods, 12 hours and 25 minutes for the channel, 11 hours and 24 minutes for the bay.[12] Both have great tidal ranges. Yet, of the two, only the Bay of Fundy has a bore. In its upper arms, Minas Basin and Chignecto Bay, the mud flats will be totally exposed at low tide, to be 40 to 50 feet under water at the flood.[13]

Other bores might be noted. In southwest England, the 30- to 40-foot tides of the restricted Bristol Channel produce Great Britain's largest, a 4-foot bore which periodically invades the Severn River. In India's province of West Bengal, the Hooghly River, the Ganges's most important channel, has a 6- to 9-foot bore, big enough to have drowned the troops of Robert Clive had the famous fording of it by that conquering English officer been made at the wrong time.[14] In Alaska, the Cook Inlet's Turnagain Arm has a bore. Only 4 to 6 feet high, it is nonetheless considered dangerous. The sound of its advance has been compared with that of surf, being audible (with an onshore wind) about 30 minutes before its arrival at a given point.

Interestingly, one does not think of even the lesser vagaries of nature's violence being periodically inflicted on La Belle France. Yet this nation, which like Switzerland is fortunate enough to be located in a belt of comparatively low seismic activity, has a bore as dangerous as it is locally famous, the Mascaret of the River

---

[12]The word *semidiurnal* means "occurring twice a day." Most of the earth's tides are of this type, there being two high and two low tides in each *lunar* day of 24 hours and 50 minutes.

[13]Off the northwest coast of France, two miles seaward from the Normandy-Brittany border, lies the famous islet and abbey of Mont-Saint-Michel. The tidal range here equals that of the Bay of Fundy, but it should be noted that these waters are part of the English Channel. There is no bore, as would be the case in a narrowly enclosed area, but simply a rapid incoming of the sea. In 1058, when Benedictine monks completed the Romanesque church and monastery, Duke William of Normandy entertained King Harold of England in the abbey. Eight years later they met at Hastings, one of the great decisive battles of history.

[14]On June 22, 1757, Clive, with less than 3,000 men in his command, crossed the Hooghly to attack an army of 60,000 under Suraj-ud-Dowlah, Nawab of Bengal. The victory of Plassey which followed, avenged the infamous "Black Hole of Calcutta" and established British power in India for nearly two centuries.

Seine.[15] The visitor to Paris might find this hard to believe. Standing on one of the Seine's bridges in the capital, or watching its quiet flow from the Île de la Cité, it would be difficult to think of this river as other than placid. Yet at Caudebec-en-Caux, a town on one of the Seine's hairpin bends between Rouen and Le Havre, the vernal bore attains a height of 24 feet and ascends the river with formidable power. Nowadays, it's a tide that runs only in the spring, dredging having restricted its operational zone and limited its season.

In our brief treatment of the subject of unusual or unfamiliar tidal forms, one other phenomenon bears mentioning, viz., the whirlpool. Students of the Greek classics have, of course, read of the legend of Scylla and Charybdis. According to mythology, Scylla, a sea numph, was rival to the enchantress Circe for the love of the sea-god Glaucus. Because the god preferred the other, Circe changed Scylla into a monster, confining her to a rock between Italy and Sicily on the mainland side of the Strait of Messina. Since the whirlpool of Charybdis was just off the Sicilian coast, the phrase was coined "between Scylla and Charybdis," referring to points of equal peril.

Most legends are founded on fact. This whirlpool actually exists and is marked on modern maps as *Scilla,* with a slight change of spelling. However, it is small. With a tidal range of about a foot and a speed of about 5 knots, its power (as observed in later ages, at any rate) is scarcely a navigational hazard.

What causes a whirlpool? Actually, they seldom develop except in areas where specific tidal conditions produce what is called a *hydraulic gradient.* This is sometimes seen when a somewhat constricted channel spans the gap between two basins. If at the ends of such a channel the tidal time sequences differ appreciably, the establishment of hydraulic gradient creates disparate water levels and the flow of the current is literally downhill. Similarly, a large difference in tidal range may produce the same effect. Such factors as bottom of coastal topography, or opposing currents or tides, then combine to induce a rotary motion.

---

[15]Although the great tourist interest it arouses has caused it to assume capitalization, *Mascaret* is simply the French word for *bore.*

Not all whirlpools are so benign as that of Scilla. The whirlpool rapids below Niagara Falls are notorious for both their great volume and extreme violence, and those below South Africa's Victoria Falls are only scarcely less so.[16] And ask any British seaman who has ever served at Scapa Flow in the Orkneys what he thinks of the eddies of Pentland Firth, a stretch of water known to local Scots as the Swilkie. In all likelihood, he'll say he wouldn't cross it in a small boat even in calm weather. With a nor'wester working up and the tide ebbing, he wouldn't cross it in *anything*. Just to see the Swilkie is unforgettable. With the tide running at the high point of the ebb, it is not one whirlpool, but many, and between their dangerous concavities the racing, very turbulent water is often weirdly domed.

Not even the briefest discussion of whirlpools can afford to omit the most famous of them all, the Norwegian Moskenstraumen, less correctly called *Maelstrom* after the short story by Edgar Allen Poe.[17] This phenomenon is very real. Running off Norway's west coast between two of the Lofoten Islands, it consists, like the eddies of the Pentland Firth, not of a single gigantic whirlpool but of many. Perhaps due partially to Poe's flamboyant hyperbole (he described how the maelstrom in full fury made the mountain tremble "to its very base"), of recent years a tendency has developed to speak of this current in terms almost of disparagement. It should not be underestimated. When the tide is at maximum flow its speed reaches a virtually unmatched 16 knots. At such times, during an interval of six hours it moves 100 million tons of water through the strait and is one of the most dangerous passages on earth.

In their publication, *Sailing Directions for the Northwest and North Coasts of Norway,* Norwegian maritime officials have this to say about it:

Though rumor has greatly exaggerated the importance of the Mal-ström, or more properly Moskenstraumen, which runs between Mos-

---

[16]The roiling of this whirlpool is said to strongly increase the turbidity of the Zambezi River for nearly half a mile downstream. The Zambezi is 5,580 feet wide at the falls, and the latter's height varies from 350 to 400 feet.

[17]In Poe's imaginative rendering, "A Descent into the Maelstrom," an old man takes a friend to a cliff above the sea, inviting him to survey the dangerous looking strait far below. While they watch, the nature of the current changes and the mighty whirlpool begins, whereupon the graybeard tells his harrowing tale.

ken and Lofotodden, it is still the most dangerous tideway in Lofoten, its violence being due, in great measure, to the irregularity of the ground. . . . As the strength of the tide increases the sea becomes heavier and the current more irregular, forming extensive eddies or whirlpools (Malström). During such periods no vessel should enter the Moskenstraumen.

These whirlpools are cavities in the form of an inverted bell, wide and rounded at the mouth and narrower toward the bottom; they are largest when first formed and are carried along with the current, diminishing gradually until they disappear; before the extinction of one, two or three more will appear, following each other like so many pits in the sea. . . . Fishermen affirm that if they are aware of their approach to a whirlpool and have time to throw an oar or any other bulky body into it they will get over it safely; the reason is that when the continuity is broken and the whirling motion of the sea interrupted by something thrown into it the water must rush suddenly in on all sides and fill up the cavity. For the same reason, in strong breezes, when the waves break, though there may be a whirling round, there can be no cavity. In the saltström boats and men have been drawn down by these vortices, and much loss of life has resulted.[18]

That seems to make it clear enough. Actually, not only small boats would be in dire peril when the current is running, but also craft of considerable tonnage. How were the directions phrased? "During such periods *no vessel* should enter the Moskenstraumen." Oceanographers have estimated that not less than 50 hydraulic currents run here at peak flow, passing simultaneously through the strait and reinforcing each other. This is comparatively shallow water. In such a vortex the power of the tide would at least equal that of the Unalga, Akutan, and Akun passes in the Aleutians, all three so treacherous the *Alaska Pilot* periodically issues special warnings about them. What effect the Norwegian current might have on a large pleasure craft, say, or even on a steamship, remains, fortunately, a matter of conjecture.

---

[18]Quoted from Rachel L. Carson, *The Sea Around Us* (New York: Oxford University Press, 1961). Note: The saltström is the most dangerous section of the Moskenstraumen.

*   *   *

In the fields of both seismology and oceanography, some writers have the habit of using the terms *seismic wave* and *seismic sea wave* indiscriminately, with the regrettable result that many readers either confuse the distinction or fail to make one at all. The difference, however, is most explicit. The nature of seismic sea waves or tsunami has already been explored at length, though we shall have more to say about them. Seismic waves, on the other hand, even though they may share with tsunami a common causal relationship in an earthquake, are quite distinctly another thing.

So far as concerns their physical effects terrestrially, seismic waves were discussed in Chapter 1, for they are simply those trains of shock waves emanating from an earthquake's focus; viz., the primary and secondary (P and S) waves and the highly destructive surface waves. The reader will recall that primary waves pass through the entire earth, readily penetrating both the outer and inner core. On the planet's far side they are not felt by ships at sea, their energy having been largely dissipated before their emergence. The side on which the epicenter is located, however, is another matter. Here, on leaving the crustal material of the ocean floor and penetrating the lesser density of sea water, this longitudinal wave, compressional in character, is bent by refraction and redirected in a nearly vertical line to the surface. Its velocity is 0.8 miles per second, or 2,880 miles per hour, slightly less than four times the speed of sound. If the shock is of relatively high magnitude, and ships are within its effective range, they will be struck by such waves. When this happens, they are sometimes said to have experienced a *seaquake*.

It is also possible for a secondary wave train to reach a ship. They are transverse rather than longitudinal, shear rather than compressional, and both slower and weaker than P waves. Passing from one medium to another, they undergo an energy reduction through partial reflection back into the crust. What remains, abruptly metamorphosed into a compressional wave, may reach the hull of a ship with sufficient force to induce its own vibrations, and will arrive along a line close to the vertical.

Quite often, the initial section of a primary wave train will arrive at the ocean's surface with insufficient strength to be detectable by a ship's personnel and will then generate an atmospheric sound wave. When such a wave reaches a frequency high enough to be audible to human ears, it will produce a series of explosive reports sometimes mistaken for thunder or salvos of heavy batteries fired at a distance. Usually, but not always, such pseudo-percussive phenomena will quickly be followed by the earliest vibrations to strike the ship's hull.

Over the centuries, the ships of all nations have repeatedly encountered seismic waves both in coastal waters and on the high seas, but only in recent times has man understood them. In the early days, entries in the logs of vessels would record a sudden shock, as though the craft had struck a reef where none was known to exist, indeed where the depth was far too great for one to exist. How could such appearing and disappearing "shoals" be explained? Obviously, they could not. But since the profession of seafaring was rife with superstition in any case, such occurrences, though feared, were simply shrugged off, dismissed as still another of those unknowns it was not given to man to fathom. The following brief descriptions make fascinating reading:

> Many interesting records have been given by ships' captains concerning the effects of earthquakes at sea. . . . In 1931 a slight shock occurred under the North Sea, and, though the water was calm, it was felt by a number of ships. Sounds were heard like distant gun fire, and under water explosions were also reported. A few minutes after the shock a heavy confused swell developed, and waves appeared to roll in from all directions. During the Lisbon earthquake of 1755 the wooden ships off St. Lucar were violently shaken, and many a captain thought he had grounded. One reported that he seemed to have struck a rock, and his ship was so strained and shaken that the deck seams opened and the compass was overturned. This shuddering effect seems characteristic. . . . In the Valparaiso earthquake of 1906 it was noticeable even at a distance of 120 miles from the centre of the disturbance, where a ship was shaken so violently that it seemed she must fall to pieces. In Valparaiso harbor it felt as though the ships were being driven rapidly through the water. . . . The same sensations were reported during the California earthquake of the same year. Ships appeared to have grounded, or struck rocks, and in some cases it seemed as though the anchor chains were running out of the hawser pipes, and the ships seemed to be dragging themselves over soft ground.[19]

[19]R. Hewitt, *From Earthquake, Fire and Flood* (New York: Charles Scribner's Sons, 1957).

In the report on the shock occurring in 1931, the reference to sounds "like distant gun fire" is clearly to the sonic boom of the primary wave. Since "under-water explosions" are not characteristic of seismic waves, these noises too were probably attributable to the same atmospheric source, for few things may be more deceptive.

The Mexican earthquake of June 3, 1932, produced interesting effects at sea. Its epicenter was located a few miles southwest of Guadalajara, in a range of 8,000- and 9,000-foot volcanic peaks which geologists speculate may be part of the Pacific's Clarion Fracture Zone. A number of ships reported their experiences. In comparing their reports, the variables due to differing locations were obvious.

During the early morning hours of June 3, 1932, the SS *Solana* was steaming through a smooth sea with light variable winds near 18°30'N., 104°08'W. At 1037 GMT she was violently shaken for about 7 sec. The ship was then about 60 mi. (170°) from the epicenter in approximately 800 fathoms of water and did not detect any change in the state of the sea. The perpendicular distance to the fault zone was also about 60 mi.

A few miles to the southwest at 18°20'N., 104°32'W., the MV *Sevenor* experienced, at the same time, vibrations that were less severe, but of longer duration (1 min.). The *Sevenor* was approximately 70 mi. (191°) from the epicenter and the perpendicular distance to the fault zone was between 65 and 70 mi. The ship reported a calm sea and slight westerly swells and detected no noticeable change in the surface of the sea.

Conditions aboard the MV *Northern Sun* at 19°56'N., 106°14'W., were entirely different. Although the vessel was 115 mi. (285°) from the epicenter, the perpendicular distance to the probable fault zone was probably not more than 10 mi. Vibrations, commencing at 1029 GMT, continued for 3 min. and became so violent that the engines were stopped. Before the earthquake, the sea had been smooth with a slight westerly swell, but by 1046 GMT the swell pattern had changed and the sea was confused.[20]

Since the term *fault zone* itself is somewhat imprecise, references to "the perpendicular distance to the fault zone" can be misleading.

[20]Quoted from Frank P. Rossi, *Mariners Weather Log,* vol. 2, no. 5 (September, 1967).

For purposes of clarification, therefore, the reader might substitute the word *focus*.

It is instructive to note that the aftershocks of this earthquake produced an effect on the sea scarcely less violent than that of the main one.

> The disturbances commenced at 0530 GMT on June 4, and the rolling continued for 15 min. Seven hours later at 1245 GMT in 19°31'N., 105°45'W. the crew aboard the SS *Talamanca* heard a loud noise like distant gunfire, then experienced severe vibrations. . . . 20 min. later the sea surface was littered for 50 or 60 mi. with small white oval objects, presumably dead fish.[21]

Reports of ships encountering seaquakes all bear a certain similarity. It has been argued that seismic waves striking a vessel's hull cannot materially damage her, a contention refuted by the following account.

> . . . on June 15, 1966, the MV *Ninghai* at 10°35'S., 161°5'E., in the Solomon Islands, reported the following damage after being shaken repeatedly at various intervals for about 2 hr. 'The Cathode ray tube shattered, the capillary tube in the barometer (was) smashed, valves were shaken out of their sockets in the wireless transmitter, the suspension wire on the gyro snapped and the azimuth mirror on the Monkey Island gyro repeater fell off. In addition we made some water in No. 3 double bottoms and after peak; also the main engine fuel line was broken and the sanitary tank on the Monkey Island was holed . . . the masts whipped about a great deal, and the funnel rattled alarmingly.'[22]

There are, of course, numerous other instances which could be cited. To date, no seismic wave of which we have record has been known to sink a ship. However, were enough rivets started and plates sprung in a noncompartmented hull such a result is not inconceivable. The *Ninghai* "made some water," obviously sus-

[21]Ibid.
[22]Ibid.

taining structural damage where it could have been serious.[23] If, in such circumstances, guys had snapped and cargo shifted, disaster might easily have followed. She also leaked fuel from a broken main line. Here the risk of fire is so obvious as to scarcely need mentioning.

Few places on earth are so thoroughly gouged and broken by the millennia-long passage of great glaciers as southeast Alaska's Inside Passage. From Dixon Entrance, near the 54th parallel, to Glacier Bay on the 59th, the view from the air is that of a long sweep of crustally deformed land divided into islands large and small, and between them, threaded with treacherous tides and often choked with ice, those sounds, straits, and channels carved by glacial action.

At the northern end of the passage is Lituya Bay, a body of water which, in at least one respect, is the most spectacular on earth, and one of the most dangerous. It is a beautiful setting, majestic in its partially frozen, primeval splendor. On its northeastern extremity, the Johns Hopkins Glacier passes between Mount Quincy Adams and Mount Lituya to round the latter in a gigantic curve and proceed to the bay's edge. On the southeast, still another river of ice feeds it, a branch of the Johns Hopkins circling round Mount Grillon. Rising between 12,726 feet (Mt. Grillon) and 13,560 feet (Mt. Quincy Adams), these peaks are as big as they are desolate, a fitting backdrop for such a body of water and part of the Fairweather Range. The top of the T at the bay's head is fissured from end to end. It forms a small but highly significant section of the Denali System, the very long, intensely active Fairweather Fault.[24]

To those who know the area, the name "Fairweather" is clearly

---

[23]The hulls of most large, modern vessels are bulkheaded off into separate watertight compartments, a method of construction pioneered in mercantile craft by Britain's Cunard and White Star Lines and their famous Clyde-built ships. The *Titanic* was one of these. In 1912, the iceberg she struck in the North Atlantic ripped a 300-foot gash in her side, flooding six of her compartments. This large percentage exceeded her design limitations; damage-control parties were helpless to check the sea. Less than three hours after the collision she went down by the head, taking a toll of 1,503 lives.

[24]See Chapter 6.

inapplicable. The weather is fitful here and mostly bad, and at all seasons fog is a curse to small craft venturing along the coast. In 1778, Captain James Cook, nearing the end of his magnificently productive career, sailed the *Resolution* in these waters on his last voyage. He did not enter Lituya Bay. Somewhat beyond it to the north, encountering clear skies for once, he saw the highest point in the range, a mountain whose summit a later generation fixed at 15,320 feet. This he named Mount Fairweather.

Cook was followed by other intrepid explorers. In 1786, Jean François de La Pérouse, low on food and water, sailed his two ships into Lituya Bay, hoping to replenish his water casks and find wild game.[25] He and his crews thoroughly explored the inlet. In his journal he described it as "perhaps the most extraordinary place in the world . . . a basin of water of a depth in the middle that could not be fathomed, bordered by peaked mountains, of an excessive height, covered with snow. . . . I never saw a breath of air ruffle the surface of this water; it is never troubled but by the fall of enormous pieces of ice . . ."[26]

Having made that discovery, another man might have weighed anchor, but for days La Pérouse stayed on. Like Cook, his entire purpose for exploring this forbidding coast was to find a practical route from the Pacific to the Atlantic along the northern shore of North America.[27] He wouldn't find it here; he saw that very early. But a lively trade was begun with the Indians, beads and metal were bartered for furs, fish was salted down. At last the time came to depart. The French captain considered the situation. Not liking the dangerous shoals at the bay's entrance, he ordered his men to take soundings. Three boats were launched. Caught unexpectedly in the power of a furious ebb tide, two of them capsized, drowning

[25]Today, Mount La Pérouse (10,750 feet), just south of Mount Grillon, and La Pérouse Glacier are named after him. This glacier is unusual in that it is one of the few in the Inside Passage which feeds directly into the Gulf of Alaska.

[26]Quoted from Bates McKee, "The Restless Edge," *Our Continent: A Natural History of North America* (Washington, D.C.: National Geographic Society, 1976).

[27]The search for the fabled "Northwest Passage," identical in purpose, began generations earlier from the opposite side of the continent. Jacques Cartier, Martin Frobisher, John Davis, Francis Drake, Henry Hudson, William Baffin, such were the names of some of those who failed to find it. Finally, in an epic voyage lasting from 1903 to 1906, the Norwegian Roald Amundsen accomplished this feat. In 1911, he would be the first human to reach the South Pole.

21 seamen. The survivors sailed away with a haunting memory.

Observing the bay's remarkable depth, La Pérouse had made a written entry. Noting the waters at the entrance were shallow, he'd courted disaster in attempting to sound them. A modern geologist might have told him something interesting. Both conditions resulted from a common cause. The Johns Hopkins Glacier (already considerably shrunken in the time of La Pérouse and marked on the early charts as "Fairweather"), had once reached to the sea at Lituya's mouth, scouring its basin very deeply. It retreated at the end of the last ice age, leaving behind a vestigial terminal moraine, all but closing the bay and denying access to ships. It was that moraine which formed the shoals. The tidal effect of very deep water and a narrowly constricted mouth produced the deadly rip.

Following the French expedition, Lituya Bay did not again appear in the annals of eighteenth-century exploration, nor did white men accord it much interest in the nineteenth. The Indians deserted it, being aware of what the pale ones didn't know. As decade followed decade, only the fish, the seals, and the birds of the sea continued to occupy its lonely, icebound reaches. The twentieth century dawned. Time passed.

It was the evening of July 9, 1958, 10:17 P.M. In these high latitudes it was only now beginning to get dark. Nonetheless, on their anchored, 40-foot fishing boat Howard Ulrich and his seven-year-old son had already turned in. They weren't alone. Inside the gravel spit at the bay's mouth, two other boats had dropped their hooks, craft about the size of their own.

At 10:18, Ulrich and the boy were awakened when the boat began to pitch with great violence. As they stumbled onto the deck there came an earsplitting crash from the direction of the bay's head, followed by a sight that neither would ever forget. Bearing down on them with terrible speed was a wall of black water at least 100 feet in height, a roaring liquid scythe whose advance scoured the flanking slopes down to the native rock. There was no time to start the engine, no time to do anything. In a half-mesmerized state of speechless dread, father and son watched the wave race toward them, saw its crest lower to about 80 feet, yet sweep with ease over the 176-foot bulk of Cenotaph Island in the bay's center.

With his pulses pounding, Ulrich felt his boat tremble. She was fighting her anchor, riding bows-on to the wave, facing the oncoming horror in the only posture that would give her any chance for life. There were scarcely seconds left for a man to pray. Then, in a vast, smothering engulfment with a roar like a thousand cataracts, the wave was on them. They soared straight upward, snapping their anchor chain. With the hull nearly vertical, the boat's bow cleaved through the curling crest. Then in a dizzying drop she fell down the wave's back side, lurching sickeningly as millions of tons of raging water suddenly flattened beneath her. They were still afloat. By an even greater miracle, neither man nor boy had been washed overboard.

The other boats fared less well. That belonging to William Swanson and his wife was carried to sea and foundered, though the remarkably fortunate pair saved themselves in a dinghy. The owners of the third boat made a costly mistake. In an attempt to out-distance the wave and clear the bay, the Orville Wagners turned and ran before it, presenting their stern to its approach. They never had a chance. Overtaken in seconds, instantly smothered and pooped, they were never seen again.

Though caused by an earthquake, the type of wave just described is not a tsunami, but what geologists and oceanographers call a *swash*. Lituya Bay is particularly subject to them. However, that which occurred in 1958, in geologic history sometimes called "the ultimate wave," established an all-time record. Investigation showed that 40 million cubic yards of rock, and ice weighing 90 million tons, had avalanched from the Fairweather Range and fallen 3,000 feet into Gilbert Inlet, the T's northwest branch. The resulting surge was incredible. Ripping and tearing its way to a height of *1,740 feet* on the opposite slope, the mighty torrent removed trees by the thousand, many with diameters of more than 4 feet, some by up-rooting them, others by snapping them off. Four square miles of forest were utterly destroyed, wiped out, a growth reaching inland for 3,600 feet from the high tide line along the shore. Then the wave killed the Wagners. In their direction, encountering only the slight obstacle of Cenotaph Island, it remained comparatively small. As it swept down the bay it left a trimline of denuded vegetation 110 feet high.

Scientists were quick to react. The morning after the earthquake, Don Miller, of the U.S. Geological Survey, flew a float plane over the bay on photographic reconnaissance. No landing was possible.

147

From one end to the other Lituya was choked with debris worse than any man-made logjam. But the investigations continued. The speed of the wave was estimated at 97 to 130 miles per hour. Then, finding for the first time the signs of earlier trimlines, they took notes to add to the geological record. At least four older waves had left their marks. By felling trees which had taken root since, and counting their growth rings, they could estimate the times of occurrence.[28]

The cause of the shock was pinpointed. Where it passes under the head of the bay it was found that the oceanic side of the Fairweather Fault had risen 3.5 feet and shifted 21.5 feet to the northwest. At Uppsala, Sweden, the long-period seismograph detected it. Its record showed that Rayleigh Mantle waves from Lituya Bay had rounded the earth six times. Seismologists were fascinated. What had Beno Gutenberg predicted? A certain type of quake could jar the planet into free vibration, he'd said, establish a wave rhythm that would set it "ringing like a bell." Once again, the great man had been vindicated.[29]

[28]The 1854 trimline reached 395 feet; that of 1874, 80 feet; that of 1899, 200 feet; that of 1936, 490 feet.

[29]The seismologist after whom the Gutenberg discontinuity was named had a disconcerting habit of being right. Predictions he made in 1926 concerning the extent and nature of the mantle were substantiated decades later by nuclear detonation testing. His theory of planetary "free vibration" received additional support resulting from the Chilean quake of May 22, 1960.

$$\sim\!\!\sim 10 \sim\!\!\sim$$

Ancient Disasters

Seismic instability of the eastern Mediterranean.   The evidence of the
spade: Heinrich Schliemann and Sir Arthur Evans.   Santorini, Crete,
and the Bronze Age destruction of the Minoan Empire.   The greatest
tsunami in the recorded history of man.

In the second section of Chapter 8, that dealing with the destruction
of Lisbon, it was stated that the northern edge of the African Plate
divides the Mediterranean latitudinally, passing through the Strait
of Gibraltar to extend all the way to the Mid-Atlantic Ridge. It was
further stated that this plate is being subducted by the Eurasian
Plate, that in the process certain Greek and Italian volcanoes were
created ages ago, among them the still active vents of Thera, Vesuvius, Etna, and Stromboli. Actually, despite its association in the
popular mind with tranquillity and scenic splendor, and the con-
juring of sleepy Italian seaports crowding sunny, infinitely romantic
beaches (all of it realistic enough), the eastern Mediterranean is and
has always been a region of profound seismic instability, the scene
of natural catastrophes of a scope not easily matched elsewhere.
   One does not have to look far to see why this is so. Not only does
plate subduction occur throughout the length of this nearly tideless
sea, but in its eastern half the great trench paralleling the elliptical
curve of the Mediterranean Ridge bodes as ill for this region as
does the Puerto Rico Trench for the Antilles or the Peru-Chile
Trench for the west coast of South America.[1] Thus, we find that
ancient folklore speaks of frightful disasters both seismic and vol-

[1]See footnote [21] in Chapter 7 dealing with this subject

canic, such as the earthquake which in 1900 B.C. devastated eastern Greece, destroying Athens, Piraeus, and Eleusis, generating tsunami which drowned the entire population of Attica.[2]

Nor are large Mediterranean earthquakes especially rare in very recent times. The following quote is revealing:

> July 9, 1956, in the Cyclades, two shocks of magnitude 7.7 and 7.2, with thirteen minutes between them. The worst damage occurred in the island of Santorin, where the famous volcano grew active again, shooting up incandescent dust and clouds of ash. The earthquake was not volcanic but tectonic, however, and it was followed by a tsunami which struck the Cyclades, reaching a height of over 80 feet on the shores of Amorgos and Astypalaea, 33 feet at Phalegandros, 13 at Patmos, and about 6 on the northeast shore of Crete. Thirty boats were sunk; 53 dead, 100 hurt, 500 houses ruined, 1,500 seriously damaged.[3]

Six feet on the northeast shore of Crete. When one reads these words, how fascinating it is to contemplate the fact that the coasts of this Aegean island, once a virtually impregnable bastion of the ancient Minoan Empire, today reveal to geologists a flooding contour *300 feet above sea level,* the record of tsunami which, together with earthquakes and volcanic eruptions, destroyed a great civilization. Obviously, this disaster was *catastrophic.* But did it perhaps transcend the usual meaning of that word, surpassing it in both the common and geological sense? Was it possibly *abnormal,* far beyond the experience of modern man, a cataclysm of an extent and destructive fury to stagger the imagination? Significantly, that 300-foot contour may also be seen along the entire Aegean coast of the Greek mainland and that of Aegean Turkey, and for hundreds of miles along both the north coast of Africa and the west coasts of the Middle East countries to the latitude of northern Syria. In the west it is found beyond Tripoli, indeed as far as Tunis at the Mediterranean's narrow waist. A few miles west of Cyrene (modern Shahhat) it disappears, only to reestablish itself dramatically across the entire sweep of the very broad delta of the Nile.

---

[2]In semilegendary form this great catastrophe has come down to us as the Ogygian Deluge.

[3]Tazieff, *When the Earth Trembles* (New York: Harcourt, Brace & World, Inc., 1964).

Clearly, the scope of the event which caused such stupendous flooding has never been equaled since. Without doubt it was the mightiest natural disaster in history, and if its peer can be found in prehistory it must surely have occurred only in antediluvian ages. But what was it precisely? What do we know about it? For answer, we must focus our attention first on that part of the Mediterranean so steeped in the lore of ancient Greece, the dark sea of many islands, of Jason and the Argonauts, the Trojan War, the wanderings of Ulysses, the victories of Greek over Persian—the Aegean Sea. We must delve into the distant past, deep into the shadowy recesses of antiquity.

Before the year 1871, the history of Greece was customarily dated from 776 B.C., marking the first Olympiad (and by some much later than that), and all that was known of what had gone before was largely considered legend or myth, the impenetrable cloak concealing whatever might have been the reality.[4] It was generally conceded that Homer had been a great writer, but if the heroes of his *Iliad* and *Odyssey* had ever really lived, why had no trace been found of them? Many had tried. None had succeeded.

In the fall of 1871 all of this changed. To the hill of Hissarlik on the Turkish-controlled Strait of the Hellespont came a German archaeologist, Heinrich Schliemann, a man imbued with the lifelong dream of finding Homer's lost city of Troy, the mighty citadel of the *Iliad*. Already wealthy by his own exertions, master at the age of 33 of 15 languages, including ancient and modern Greek, the scholar combined in his nature a curious mixture of practical businessman, cool adventurer, and flamboyant romanticist. He found the site of Troy, found it about where he said he would, though his selection of the second stratum instead of the correct and much shallower sixth was in error.[5] As a result, the vistas of archaeology

[4]Publishing his masterly history of Greece in 1846, George Grote observed: "I begin the real history of Greece with the first recorded Olympiad, or 776 B.C. . . . for the truth is, that historical records, properly so called, do not begin until long after this date."

[5]Characteristically, having discovered two exquisite gold diadems among a veritable treasure of gold, believing them to be part of Priam's horde, he placed them on the head of his young Greek wife, certain they had once adorned the brows of Helen of Troy. Later, in shaft Grave V at Mycenae, finding the marvelously intact body of a man in golden armor, he removed its gold mask, then wrote to the King of Greece: "I have gazed on the face of Agamemnon."

were enormously expanded and history itself given a new dimension. For Homer's heroes *did* live. Troy *did* exist and *was* beseiged by the Greeks. Today, most historians accept one of two possible dates for the city's fall: 1184 B.C., the date claimed by the third-century B.C. Greek astronomer, Eratosthenes; or 1260 B.C., that assigned by such modern investigators as the American archaeologist Carl W. Blegen.[6] Schliemann's discovery had expanded the history of Greece by a period of centuries. His subsequent digs at Mycenae, Ithaca, Orchomenus, and Tiryns threw even more light on what had been dark for so long.

The other principal figure in the story of this "revolution in archaeology" was a brilliant Englishman, a man as dedicated as Schliemann himself and as enterprising, Sir Arthur Evans. Taking his inspiration from the German, building on Schliemann's foundation in his own dogged pursuit of knowledge, in the spring of 1899 he was on the island of Crete, digging at the mound of Kephala near modern Heraklion, the legendary site of ancient Knossos. Like Schliemann, he was successful, but what he discovered far exceeded his expectations. Having found in Athens some small stones bearing what appeared to be hieroglyphic symbols, having been told they came from Crete, he was there in the hope of deciphering what might prove to be the earliest system of European writing.[7] Instead, he uncovered a vanished civilization.

Knossos, whose great palace Evans excavated, was the capital city of a people who once ruled an island empire in the Aegean. Although their area of immediate control included only Crete, the group of islands called the Cyclades and certain regions of the Greek mainland, their fleet was so powerful and their sea trade so extensive that their influence spread from end to end of the Mediterranean, its reach embracing all the lands of that basin from the Levant

---

[6]Perhaps more than any other "post-Schliemann" expedition, that sponsored by the University of Cincinnati and led by Blegen was responsible for identifying the sixth stratum as that of Troy.

[7]Evans was correct in his guess. Subsequently, Michael Ventris and John Chadwick determined that at least two types of script were used in Bronze Age Crete, forms labeled Linear A and Linear B. To date, Linear A has kept its long-forgotten secret. Linear B, however, when deciphered, proved to be a primitive syllabic Greek dating from about 1400 B.C.

to Tunisia, and beyond to the British Isles and even to Denmark.

This race, named by Evans the Minoans after their first king, Minos, exhibited characteristics which have caused some modern historians to believe that originally they may have come from Asia. If they did so, the further probability exists that they entered the Mediterranean through Anatolia, having been in extensive contact there with both the Hittites and the Hurrians. By 2000 B.C. most of their 90 cities had been built on Crete, including Knossos and its palace, and the palaces of Mallia, Phaistos, and Hagia Triada. Theirs was a Bronze Age civilization, contemporary with that of Mycenae on the Greek mainland; Phoenicia with its famous ports of Tyre and Sidon; Sumer followed by Babylon in Mesopotamia; the Old, Middle, and New Kingdoms of Egypt; Mohenjo Daro and Harappa in the Indus Valley; and cultures existing in China from the last decades of the legendary Hsia Period to the decline of the Shang Dynasty.

To a very large degree, the great period of all Bronze Age civilizations was spanned by the second millennium, i.e., from 2000 to 1000 B.C. In the case of the Minoans, a peaceful people in earlier centuries, their absolute command of the sea allowed their industries and arts to flourish unabated, as often occurs when the energy of a race is not sapped by the rigors and exigencies of war. They were gay and fun-loving, endowed with an almost ethereal sense of beauty. Although their artifacts reveal a religion evolving from nature worship through a mother goddess, the sacrifice of bulls, and the ultimate emergence of a divine queen and her consort, they were never priest-ridden in the sense of their contemporaries, thus were spared the stultifying repression and debasing cruelty so often the bane of enlightenment and progress.[8] Indeed, they were far advanced for their time. Until about 1600 B.C., they remained virtually unchallenged in the eastern Mediterranean. After that there rose in rivalry on the mainland of Greece a nation of

---

[8]Unlike so many ancient peoples, the Cretans did not erect temples but worshiped instead at many consecrated sites, often wilderness shrines raised in forests, mountain retreats, or large caves. Sacred areas were usually identified by such religious symbols as the carved horns of a bull and the Cretan double axe, a weapon probably used in the sacrifice of the animals.

Aryan invaders who had replaced the native Neolithic culture by force of arms, the Mycenaeans.

What became of the Minoans? The exact date of their vanishing from history remains even today a subject for argument and conjecture. However, since their large and efficient mercantile marine did a thriving trade with Egypt, it is not remarkable that Egyptian scribes recorded the abrupt cessation of that trade and the utter disappearance from the Mediterranean of the Minoan merchant fleet. This was a fantastic occurrence. Yet the Egyptians knew the nature of this disaster, if not its precise source, for they had shared in it, though at a distance sufficient to be spared annihilation. Somewhere to the west an island had exploded. They understood that much. Following this, the sky had turned black as the sun disappeared. The streams and rivers had been clogged with ashes, the sea with pumice, and great waves had risen to fall upon the land . . .

In the quotation from Haroun Tazieff cited above, the word *Santorin* was mentioned. More usually called Santorini, it is a medieval Italian corruption of *Santa Irena*, a saint who was martyred on the island of Thera and remains today its patroness. Lying in the south-central Aegean, 138 miles southeast of Athens and 78 miles north-northwest of the island of Crete, Santorini is the name given, collectively, to a broken group of five islands, two partially encircling and three within a gigantic, ocean-filled caldera. By far the largest of these is Thera, which forms about two-thirds of the outer ring, roughly an ellipse on the north, east, and south. Second, and much smaller, is Therasia, a part of the circle in the northwest quadrant, while the tiny island of Aspronisi (white island) lies within the larger of the ring's two openings, the broad channel in the southwest.

Reminiscent of both Anak Krakatoa in its lagoon, and of Wizard Island, vent plug of Oregon's Mount Mazama in the caldera waters of Crater Lake, are the two islands almost in the center of the Santorini complex. The larger, Nea Kameni (meaning "the new burned one" in Greek), broke water in stages between 1707 and 1711, and its bulk was considerably added to by eruptions occurring between 1866 and 1870. The smaller, Palaea Kameni ("ancient burned one"), about one-sixth the size of its companion, lies immediately southwest of it with only a narrow channel between. It first appeared above water in 198 B.C. Since then it has erupted five times, each outbreak increasing its size until its present height of 300 feet was reached. Once there was a third island in the bay, Mikra Kameni ("little burned one"), which appeared in 1573. Now it is part of Nea Kameni, the result of lava extruded from 1925 to 1929.

In addition to the volcanic activity of 1956, previously noted, there have been two other recent eruptions, in 1939 and 1950, neither greatly altering the contours of the central islands. Yet the volcano remains active. Obviously, in time, Nea Kameni and Palaea Kameni are destined to become one, as once they were long ago.

The visitor from Greece will normally enter the bay through the northwestern channel between Therasia and Thera. When he does so he will feel that the ship has suddenly become very small, that he himself is reduced to the minutest of specks in the universe. Everywhere beneath the vessel's keel the water runs to a depth of between 300 and 400 meters (984 to 1,312 feet), making it impossible to anchor. He is struck by the size of the basin, 7 miles long from north to south, 4.66 miles wide from east to west, enclosing an area of 32 square miles. Perhaps even more than the ominous masses of black lava in the center of the bay, the seemingly endless cliffs on his flanks inspire awe. There are no beaches. Rising sheer from the water, the escarpments average 1,000 feet in height, reaching 1,150 feet in some places, and are banded in curious striations of grayish black, dirty white, and, forming the lowest mass and running two-thirds of the way up at some points, a deep, very startling red somewhere between the color of brick and a certain shade of porphyry found only in Finland's Karelian Isthmus.[9]

Could anyone possibly live in such a setting? Here and there along the upper edges are man-made structures all in gleaming white, so white that the sun's reflection tends to make them blind-

---

[9]This porphyry is an incredible red-purple, the exact shade defying description. Today, in Les Invalides in Paris, a large rectangular block of it surmounts the sarcophagus of Napoleon.

ing.[10] Man does persist in this bleak desolation. Each of several towns is occupied. Phera, most notable as the clifftop capital of Thera, is reached by donkeys who climb 670 feet up a zigzag stone roadway of 1,200 steps. At the bottom is a quay of stone and cement where small boats tie up, many engaged in fishing and octopus hunting. There, too, is the bayside village of Skala, snow white like all the others.

Momentarily, the newly arrived traveler experiences a sense of relief. After all, the place is not devoid of life. In some ways, these precariously perched settlements resemble other Greek coastal towns. Then, inevitably, his imagination soars. Everywhere he looks is grim evidence of a natural convulsion of indescribable scope and power, a cataclysm on the order of that which must have witnessed the creation of the world. The visitor recoils. His mind cannot encompass it. The power that blasted this great hole, the resultant titanic holocaust of fire, the effect on the surrounding sea, now so benign. Though he may not be a scientist, he'll instinctively realize that here is something unique. He may travel all his life, cross every continent, navigate every ocean; nowhere will he find again what he sees here. It is the most colossal caldera on earth. Before him is visual evidence of an eruption which in the distant past doomed an entire civilization to extinction. That civilization was the Minoan Empire.

In the second half of the nineteenth century, two French expeditions to Santorini stimulated interest in geological circles. In 1866, the vulcanologist Ferdinand Fouqué carried out excavations on Therasia.[11] In 1870, his work was continued by Mamet and Gorceix with their digs on Therasia and Thera. Their successive findings were fascinating. Clearly, the broken ring of islets had once been a single island of quite large dimensions, its mountainous center having collapsed long ago as the result of an incredibly violent eruption. The signs of ancient human habitation were plentiful. Digging through heavy layers of pumice, the scientists found

---

[10]Typically, throughout those islands of the Cyclades and Dodecanese groups which are inhabited, the homes of Greeks are of whitewashed stone, witness to the fact that many centuries ago they knew that white reflects heat superbly.

[11]Due to the demand for cement to be used in the construction of the French-built Suez Canal, mining of Theran pumice had been extensive. It was the inadvertent discovery of Bronze Age artifacts which alerted Fouqué.

both copper and stone tools, mortars, pestles, and millstones fashioned from lava, decorated pottery, and even limestone houses with painted frescoes.[12] Gradually, the evidence began to accumulate of an advanced Bronze Age culture superimposed on the neolithic, a people capable of creating sophisticated art and architecture, possessing a numbering system and a system of weights and measures. They were skilled agriculturists. Their pottery and basket-weaving showed a high level of development.

Even at this early stage of exploration there was some evidence to suggest a possible connection between Minoan Crete and the civilization that had flourished on Santorini, and more than one scholar began to think along these lines. It was not until 1967, however, that the question was resolved. In that year, Professor Spyridon Marinatos of the Archaeological Society of Athens conducted excavations at Akrotiri in the southwestern region of Thera. Nine trenches were dug. Among the many artifacts recovered were painted storage vessels or *pithoi* identified as dating from the Late Minoan I Period. Among the architectural remains were a stone wall of this period and the stone façade of a Minoan villa or palace. Fresco fragments revealed glimpses of ancient art all in the Minoan style.

As the excavations were continued the following year, more and more proof came to light. An entire Minoan town was unearthed. With two notable exceptions, the absence of human remains seemed to indicate that the terrible disaster which overwhelmed the island was sufficiently gradual to permit evacuation, perhaps to Anaphi, Ios, or Melos, perhaps to more distant Crete. It is doubtful that those who may have left in the final stages of the eruption ever reached their destinations. The ash fall would have been continuous and devastating, floating pumice a possibly insuperable barrier. Those who escaped asphyxiation or drowning might easily have been burned to death.[13]

In 1939, Marinatos published his theory that the eruption and

---

[12]Some of the houses found by Fouqué had been broken in two with only part remaining to be seen, the result of the shearing effect when the volcano's summit abruptly subsided.

[13]At Phera, in 1956, some teeth belonging to a man and a woman were found in a *pozzolana*, or pumice mine. Two of them had been charred by intense heat.

disintegration of Thera during the Bronze Age had utterly devastated Crete and resulted in the fall of the Minoan Empire, a belief which at the time was shared by very few. Disagreeing with those who attributed this catastrophe to an earthquake, he wrote: "A normal earthquake is wholly insufficient to explain so great a disaster. In all the many earthquakes known to us, there has never been such a widespread destruction at one and the same moment."[14]

Commenting on the belief of the ancient Egyptians that Crete had been totally submerged, in 1950 he wrote: "The Egyptians unquestionably had heard about the sinking of an island, which was Thera, but this island, small and insignificant, was unknown to them. This event they transferred to the neighboring Crete, an island which was dreadfully struck and with which they had lost contact suddenly."[15]

Once it became evident that a sizable population had existed on Santorini, a culturally advanced people of great antiquity, it then was clearly in the interest of science to determine as accurately as possible the date of their destruction or dispersion. There are several methods for calculating this. Investigators are prone to use them all so that one may serve as a check against another. Among laboratory techniques, these methods include the radiocarbon process of isotopic dating and that of thermoluminescence. Both are highly useful under certain conditions. However, neither is as accurate as the chronological classification of artifacts by style and strata identification, nor as reliable as certain historical evidence amenable to cross-reference.

Since the Minoans are known to have pursued a centuries-long commerce with Egypt, and since, moreover, from the Egyptian Eighteenth Dynasty to the Persian invasion of Egypt in 525 B.C., there is a completely documented historical record, through Egyptian relative chronology (and that of such contemporary Aegean powers as the Mycenaeans) it becomes possible to establish a prob-

[14]Spyridon Marinatos, "The Volcanic Destruction of Minoan Crete," *Antiquity,* vol. 13 (Athens, 1939).

[15]Marinatos, "About the Rumour of Atlantis," *Cretika Chronika,* vol. 4 (Heraklion, 1950).

ably correct chronology for the Minoan Empire. When, in combination with laboratory analysis, this technique was applied to the inclusive dates of the last cataclysmic Thera eruption, one investigator, the American oceanographer James Mavor, was led to base his conclusions on several events.[16] Absolute chronology was established by the radiocarbon dating of wood samples found in Phera, and in ash deposits found in deep-sea cores. Supportive evidence was then supplied by relative chronology. This consisted of pottery discovered and classified during the Mamet-Gorceix excavations on Thera, and those of 1967, and the discovery of a Minoan sword whose design was contemporary with certain weapons unearthed in Mycenaean shaft graves.

Mavor has this to say:

The date of the beginning of this period of volcanic and tectonic activity is most accurately given at present by the pottery style (1500 B.C.). The duration of the period is best given by the evidence of destruction on Crete, which occurs as tsunami destruction of coastal settlements and earthquake destruction of structures all over the eastern half of Crete. Agricultural destruction due to ash fall is indicated by ash in deep-sea cores and on Thera. . . .

Galanopoulos first provided the explanations for the time gap between the beginning of the Thera eruption and the Cretan destruction.[17] He wrote that the eruption occurred in interrupted phases over a period of several decades with the *coup de grace* coming at the end of the collapse and the tsunami. His theory that earthquakes between Crete and Thera triggered the eruption were [sic] confirmed by my observations in 1967.[18]

[16]Mavor, who worked with Marinatos on Thera in 1966 and 1967, is employed by the Woods Hole Oceanographic Institution in Massachusetts. An amateur antiquarian and archaeologist, he is one of the engineers who designed the deep-sea research submarine *Alvin,* the craft which located the lost H-bomb off the Spanish coast. He was also influential in convincing the Woods Hole staff of the advisability of sending the research vessel *Chain* to Santorini.

[17]The reference is to Dr. Angelos Galanopoulos, lecturer in seismology at the University of Athens, and director of the Seismological Institute at Athens Observatory. Member for Greece of the International Union of Geodesy and Geophysics, Galanopoulos is also President of the European Seismological Commission's Subcommission on Tsunami.

[18]James W. Mavor, Jr., *Voyage to Atlantis* (New York: G. P. Putnam's Sons, 1969).

According to the calculations of Galanopoulos and Mavor, the final blow, the earth-shaking cataclysmic blast which ended the Minoan world at one stroke, could have come as late as 1450 B.C., or slightly later. As will be seen in Chapter 11, there is reason to note an hypothesis that it fell in 1447 B.C., the year marking the death of the Pharaoh Thutmose III.

In Chapter 2, the theory of convection currents in the earth's mantle was briefly noted. In reference to it, mention was made of Bruce Heezen, one of the marine geologists at Columbia's Lamont Observatory. Together with his colleague, Dragoslav Ninkovich, Heezen's involvement with Santorini began when the two scientists studied data on eastern Mediterranean cores taken during the Swedish *Albatross* expedition of 1947–48. A decade later, when Lamont's own *Vema* returned from that sea with more cores taken in the same area, Ninkovich and Heezen studied them directly for signs of the ash layer earlier reported by Greek geologists. Seven of the Swedish cores and fourteen of Lamont's revealed stratified volcanic ash in sediments of globigerina ooze. Gradually, from the work of these men and others, notably Galanopoulos, a geographic pattern emerged. Laid down in a northwest-southeast configuration, describing a gigantic oblong shape, the largely submerged ash layer covered an incredible 80,000 square miles. Running from a point just east of the Attic Peninsula, embracing nearly all of the Cyclades and Dodecanese Islands, including the eastern three-fourths of Crete, it extended nearly to the latitude of Alexandria, Egypt, some 400 miles southeast of Santorini.

It wasn't until 1965 that Heezen and Ninkovich published their data. They explained that as a result of chemical treatment of the core samples it became possible to test for the refractivity of the volcanic glass in the ash and to compare it with that in the ash found on Thera. The two were identical. Both had come from the same eruption.

In 1966, the Paris-based American science writer Daniel Behrman talked with Ninkovich, at that time still at work on the Theran investigation. When asked about his activity, the geologist replied that he was attempting to estimate the energy released by the second millennium Santorini eruption. "We think it must have been around a thousand megatons," he told Behrman. "It was at least ten times as powerful as Krakatau in 1883, the biggest known eruption in history."[19]

Numerous comparisons have been made between Krakatoa and Thera. It is both usual and logical to quote statistics revealing the enormous power of the former, then to place it in perspective by doing the same for Santorini. Thus, for example, note the following quotation from Galanopoulos and Bacon:

> Let us now compare the result of two great volcanic eruptions—that of Krakatoa in A.D. 1883, and that of Santorin in the fifteenth century B.C. The Santorin caldera is 32 sq. miles (83 sq. km) in surface and 160–220 fathoms (300–400 m) deep. The Krakatoa caldera does not exceed 8 sq. miles (22 sq. km) in area and 110–160 fathoms (200–300 m) in depth. The Santorin caldera is therefore about five times greater in volume than that of Krakatoa. The thickness of the ash thrown out at Krakatoa does not exceed 16 in. (40 cm). The ash-covered area at Santorin is much greater and the thickness of the layer reaches 100–130 feet (30–40 m).
>
> The thermal energy produced at the Santorin eruption, according to Peter Hedervari, was about three times that of the corresponding energy of Krakatoa.[20]

After citing figures to show that "the energy liberated by the eruption of Krakatau was far greater than that of any volcanic eruption in historical time," Behrman adds that nonetheless it was "insignificant" when compared with Santorini. Mavor says substantially the same thing. However, he makes it obvious that the great difference in the amount of ash blown out by the two eruptions was only partially attributable to the difference in power, noting that the continuing activity of Santorini may have lasted 30 years or more with the ash gradually accumulating. Describing the final

[19]Quoted from Daniel Behrman, *The New World of the Oceans* (Boston and Toronto: Little, Brown & Company, 1969). Note: In transliterating from the Indonesian, *Krakatau* is simply a variant of *Krakatoa*.

[20]A. G. Galanopoulos and Edward Bacon, *Atlantis—The Truth Behind the Legend* (New York: Bobbs-Merrill, 1969). Note: Among other important theses, Hedervari is author of "Volcanophysical Investigations Concerning the Energetics of the Minoan Eruption of Volcano Santorin," *Bulletin Volcanologique*, vol. 25 (Paris, 1967). Edward Bacon, author of several outstanding books on archaeology, is a distinguished British scholar who since 1945 has been archaeological editor of *The Illustrated London News*.

stages, he refers to "the paroxysmal fall of ash . . . that turned day into night."

Our subject is seismic sea waves. No matter how exhaustive his research, nowhere in scientific literature will the reader find reference to tsunami that remotely compare in size and power with those which followed Thera's climactic, self-eviscerating eruption. Taking the average depth of the eastern Mediterranean and applying the standard formula, Heezen and Ninkovich estimated the speed of the waves at 215 miles per hour. Using the Greek word *stronghyli*, meaning "round," the geological profession's pre-eruption name for Thera, they wrote as follows: "The whole northern coast of Crete must have been inundated twenty or thirty minutes after the collapse of Stronghyli Island. The Minoan tsunami may also have been destructive beyond the Aegean Sea. The most vulnerable areas were the coasts of Tunisia and the low delta lands of the Nile. The coast from Tunisia to Syria must have been inundated no more than three hours after the collapse of Stronghyli Island."[21]

Elaborating with their customary exactitude, Galanopoulos and Bacon presented a somewhat detailed picture.

A faint idea of the destructive power of the gigantic waves which followed the collapse of the central part of Stronghyle may be reached by a calculation of their height. The amplitude of a tsunami is proportionate to the initial amplitude and inversely proportionate to the square root of the distance the wave has traveled. In addition to the loss of amplitude due to the wave's spreading out, there is also a certain loss from absorption due to frictional dissipation of the elastic energy into heat. Assuming an eustatic rise of the sea-level by 6 feet or 9 feet in the time interval of about 3,500 years, the height of the sea waves in Jaffa-Tel Aviv should have been at least 22 ft.[22] Disregarding now the loss from absorption and taking into account that the sea waves produced by the collapse of Stronghyle had a height of more than 22 ft (7 m) when they reached Jaffa-Tel Aviv, about 562 miles (900 km) away, we can easily calculate that their height at starting point would be at least 690 ft (210 m). This calculation

[21]Dragoslav Ninkovich and Bruce Heezen, "Santorini Tephra," *Colston Papers*, vol. 17, (London: Butterworth Scientific Publications, 1965).

[22]Webster defines *eustatic* as "relating to or characterized by worldwide change of sea level."

of the initial height of the tsunami is in accordance with Professor Marinos' recent estimate.

Interesting and illuminating effects have been found on the island of Anaphi, which lies about 15 miles (24 km) east of Santorin. Here Professor Marinos and his collaborator, N. Melidonis, found layers of pumice in three places. On the western side (that facing Santorin) they found a layer of thick-grained pumice mixed with pieces of 4–6 in. (10–15 cm) diameter; and this was 385 yds (350 m) from the coast and between 130 and 170 ft (40–50 m) above sea-level. The two other deposits are on the northeast side of the island, the side away from Santorin. The first of these is at a height of 530 ft (160 m) above sea-level and 820 yds (750 m) from the sea; while the second is about a mile (1,650 m) from the coast and 820 ft (250 m) above sea-level. All three layers of pumice were found at the heads of ravines or valleys.

According to Marinos and Melidonis, this pumice was deposited by the sea waves which followed the formation of the volcanic cavity of Santorin.[23] . . . The fact that the pumice was found at a greater height on the northeast coast of Anaphi than on the western, Santorin-facing coast can be accounted for by the eastern barrier of the island of Thera, the northwest direction of the deepest channel through which the water spread out from the caldera, and the convergence of the waves sweeping round the island. A parallel to this can be found in the tsunami which started from Chile on 23 May 1960. This reached a height of 17 ft on the southeastern side of Hawaii— the coast, that is, which directly faced the direction of the tsunami; whereas, on the northeastern side, near the town of Hilo, it reached a height of 35 ft above sea-level, and even on the northwestern shores touched the height of 20 ft.

Mavor's figures closely approximated those of Galanopoulos and Bacon. Having cited the example of the 1958 Lituya Bay swash for comparative purposes, using Don Miller's computations in part, he pointed out that the Lituya avalanche released energy on the order of 3.5×10^{14} foot-pounds, 2 percent of which went into the wave itself.[24] Aware of the similarity in fluid dynamics between a swash and a seismic sea wave, he then applied these figures to the Santorini tsunami. He wrote in *Voyage to Atlantis:*

---

[23]G. Marinos and N. Melidonis are the authors of the paper "Über Die Grösse Des Beim Vorgeschtlichen Ausbruch des Santorin-Vulkans Ausgelösten Tsunamis," *Bull. Griech. Geol. Ges.,* vol. 4 (Athens, 1959–61).

[24]See Chapter 9.

The energy of the Thera collapse . . . was 2,000 times greater than that of the Lituya Bay slide. If we accept the 2 percent of the land collapse energy as contributing to sea waves, the 7×10^{17} foot-pounds figure for the collapse energy at Thera is convertible to three ocean waves each having a height of 700 feet at Thera and a length of 80 miles. Or, in another form, we can picture three great waves each having a height at sea of 200 feet just before crashing on the Cretan shore, sweeping all humanity before them and rushing miles inland to climb hundreds of feet up the mountainside before exhausting their gigantic energies.

A height at sea of 200 feet just before crashing on the Cretan shore. Lending themselves to an unfortunate ambiguity in this case are the words "at sea," which could be construed to mean *on the open sea.* Mr. Mavor is an oceanographer. Since it must be presumed he is cognizant of the kinetic nature of tsunami, the key phrase would appear to be ". . . just before crashing on the Cretan shore . . . ," i.e., in the final minutes as the waves lifted to the suddenly encountered incline of the island's offshore slope. Perhaps the issue should not have been raised. In any case, Mavor seems to resolve it for us on another page when he refers to the waves as ". . . rushing *unseen* across the Sea of Crete. . . ."[25]

For the sake of argument, considering the undoubted fantastic height of the waves at their point of genesis and the comparatively short distances involved, it is *possible* that we might here be dealing with an utterly unique occurrence, a freak of nature never repeated in subsequent ages, i.e., a 200-foot wave *on the open sea.* However, it seems highly doubtful. From the south rim of the caldera of Thera to the nearest point on the north coast of Crete, Akra Dia (Cape Dia), is exactly 78 miles. Even in the case of such a colossus as a 700-foot wave it seems obvious that the force of gravity would flatten it out long before it traversed such a distance. One thing is certain, however. By the time they rose against the coast of Crete these tsunami were once again enormous, nightmarish. Based on archaeological evidence, the consensus is that it was tsunami, not earthquakes, which destroyed the northern coastal settlements of

---

[25]The italics are the author's.

Gournia, Mallia, Amnisos (the port of Knossos), and Nirou Khani. In his paper of 1939 Marinatos noted:

> I think there is little reason to doubt that the devastation of the coast sites of Minoan Crete was caused by the waves from the eruptions of Thera. We read of similar phenomena on the occasion of later eruptions of the same volcano, which in every case were less violent than the one in 1500 B.C. Philostratus, for example, reports that during the eruption of A.D. 60, when a new island was formed in the center, the sea receded about ¾ mile from the south coast of Crete. . . . In the eruption of A.D. 1650 in Thera itself the waves came two miles inland and swept away old walls and chapels, foundations and all.

To the trained eye, the signs of tsunami damage on the north coast of Crete are still evident today after nearly 35 centuries. For example, when Marinatos excavated the royal villa at Amnisos he noted that some of the great stone blocks in the ruin of the west wall had been driven *outward,* while the south wall showed gaps where others had simply vanished. He reasoned that an earthquake could not have done this. In the case of the missing blocks, especially, it seemed obvious that they had first been thrown down, then swept out to sea in the mighty backwash of the waves.

One of the most interesting phenomena associated with tsunami, and still largely in the realm of the theoretical, is the hypothesis that under certain conditions of concomitant reasonance and speed volcanically generated air vibrations can not only reinforce and amplify seismic sea waves, but can actually *re-excite* such waves whose pattern and thrust have been broken by an island barrier—thus, in effect, regenerating the tsunami on the island's opposite side.

Speaking of the Krakatoa eruption, Behrman makes a single brief reference to such concussive displacements of air, as follows: "In a short period of time, tsunami were recorded on the coasts of nearly all oceans. The aerial vibrations produced tremendous roars or destructive vibrations, depending on the length of the waves. The area where the roar was heard is about one-thirteenth of the surface of the globe."

D. S. Halacy also touches on it: "Some researchers believe that the great tsunami that followed the eruption of Krakatoa was caused not by motion of the ocean bottom, but by the air blast from the volcano. This theory is not universally accepted by scientists, but it is generally agreed that tsunami do produce atmospheric waves. . . ."[26]

Mavor elaborates at greater length. On the Thera eruption he writes:

There is no doubt that the waves hit the African coast. Two men from the Massachusetts Institute of Technology, Frank Press and David Harkrider, had recently come to the startling conclusion that sea waves created by the Krakatoa collapse were amplified by air blast waves which happened to be in synchronism with them. This means that sea waves are actually able to break on land, and then, propagated overland through the air, can be reestablished and again agitate the sea beyond. This phenomenon, to be sure, requires special but by no means improbable conditions, that is, an air blast of the same velocity as the sea waves. This resonance is a function of the ocean depth and the magnitude of the blast.

I applied their results to the Thera problem and found that Crete need not have been an obstacle to propagation of the waves that we know came from Thera. The waves, but not the water, of course, could have literally jumped over Crete, transmitted by air vibrations which would then build up the sea on the other side.

Galanopoulos and Bacon agree in principle. They observe:

The atmospheric waves generated by the Krakatoa eruption jumped over the land barriers and reexcited sea waves strong enough to break anchor chains of ships moored in the Port of Valparaiso, Chile. It was estimated "that a surface explosion amounting to about 100–150 megatons would produce pressure pulses equivalent to those observed from Krakatoa."

As Halacy has remarked, in scientific circles this theory has not gained anything like complete acceptance, and to date remains entirely unproved. That titanic explosions create very powerful atmospheric waves is quite well known. But can such masses of

[26]D. S. Halacy, *Earthquakes: A Natural History* (New York: Bobbs-Merrill Company, 1974).

suddenly displaced air actually reactivate tsunami whose force has already been expended against a coastline? Mavor writes that the Theran waves hit the African coast. This is certainly true enough, but they need not have been "reactivated" south of Crete to do so. Already mentioned was the 300-foot contour line still visible today. In the west it extends from Tunis to Cyrene, and in the east from Matrûh to the northward-bearing coasts of the Middle East countries. Between the two points, i.e., from Cyrene to Matrûh, it is not to be found—and that is precisely that stretch of the African coast which lies to the south of Crete! Apparently, the island *did* act as a barrier, and a very positive one. About 160 miles long in its east-west axis, 6 to 35 miles wide, it is mountainous literally from end to end, the highest peak in the range, snow-crowned Mount Ida, reaching 8,878 feet.[27] A formidable obstacle, it would seem. Nonetheless, the theory of tsunami repropagation by air waves remains an intriguing question, one of the more challenging areas for future research.

There is a final point to briefly explore, that of the *nature* of the Santorini eruption. As the reader will recall from Chapter 3, the thing which gave Krakatoa its incredible power was not the collapse of the summit in itself, but the fact that such titanic subsidence admitted a vast quantity of sea water into the subterranean magma chamber, thus creating steam pressure of incalculable intensity. The result was terrific explosions. The fallen summit of the mountain (actually the collapsed roof of the magma chamber) was then blown out again, leaving the imposing caldera we see today.

All available evidence indicates that Santorini was the same. There, too, it was water mixing with incandescent magma that generated steam, but a larger crustal lesion, a greater liquid infusion, a far greater pressure of expanding gas than at Krakatoa. The nature of the ultimate cataclysm was monstrously explosive. Yet, curiously, many modern investigators tend to ignore this factor. They write of eruptive exhaustion of the products of the magma chamber, its collapse due to lack of support, the inrush of the waters of the sea, the water bursting out again to form tsunami. They seldom mention *steam.*

[27]According to Greek mythology, Mount Ida was the birthplace of Zeus, paramount deity of the Olympian gods.

Let the reader make no mistake. Steam pressure produced the culminating detonations at Krakatoa, and that last earth-shattering blast at Thera. In all the world there has never been so destructive a force as constantly building steam pressure under close, rockbound confinement. Almost certainly, there never will be. To see the power of this force exceeded within our solar system we must look to our sun, to solar prominences (hydrogen explosions) often greater than 100,000 miles high.

11

The Santorini Tsunami

The Hebrew Exodus from Egypt.

Prominent among those scholars who uphold the theory that Santorini is actually the lost and legendary Atlantis are Galanopoulos, Bacon, and Mavor. They have written extraordinarily erudite and meticulously researched books on the subject, two little-known works that were published in 1969. Professor Marinatos, on the other hand, regardless of what his private conviction may have been, for decades avoided lending public support to this theory, repudiating it to the press in Athens as late as 1966. It is the opinion of the present writer that a desire to avoid ridicule motivated this distinguished scientist, the subject being still at that time extremely controversial. Yet in 1967 Marinatos altered his position. Interviewed by an Italian magazine, urged to express himself freely on the subject of Santorini, he at first seemed to deal in somewhat vague generalities, then was faced with the following direct question:

"You are convinced that this is Atlantis?"

The reply was unequivocal. "I am convinced. . . ."[1]

Although for a very long time it has been one of history's most fascinating puzzles, it is not part of this book's purpose to deal with Atlantis except to mention it in passing as a corollary to the destruction of Minoan Crete, a destruction involving the greatest tsunami to be found in the human record. Scores of volumes have been written on this heretofore legendary civilization, the vast ma-

[1]Mavor, *Voyage to Atlantis* (New York: C. P. Putnam's Sons, 1969).

jority of them undertaken without sound scientific reasoning, with the layman's passion for expounding theories not amenable to controlled research or experimentation. In their fascinating study, Galanopoulos and Bacon utilize considerable space demolishing hypotheses standing counter to their own, and do so with comparative ease and assurance. They are ably seconded by Mavor. In fact, it is difficult to read either of these mutually supporting accounts without giving serious consideration to Santorini as the most logical site for Atlantis, sharing with Crete's capital city of Knossos the central power position within the Atlantean Empire. The evidence presented is abundant and most convincing, and in numerous instances backed by provable fact. Both books are highly recommended.[2]

One very intriguing aspect of the Bronze-Age Santorini eruption and its resultant tsunami is the possibility that it may have been directly connected with certain ancient traditions of the Flood. References to a widespread and frightful inundation are by no means restricted to the Book of Genesis of the Christian Bible. Greek writings, for example, tell of Deucalion's Flood, a rising of the sea great enough to drown much of the mainland of Greece, the Aegean Islands, and most of the coastline of the eastern Mediterranean. At Ugarit (modern Ras Shamra) a tablet was discovered which spoke of that port's destruction by enormous waves some time between 1450 and 1400 B.C., while another recorded a second annihilating strike in the same area about 1370 B.C.

Readers of the great archaeological classics of the nineteenth century will recall the works of the Englishman Austen Henry Layard, particularly the accounts of his excavations at Nimrud and Nineveh in what is today Iraq, and his discoveries of the palaces of Assurnasirpal II, Sennacherib, and Assurbanipal. Among the priceless cuneiform tablets which Layard gave to the British Mu-

---

[2]At this writing, there is a consensus in international scientific circles favoring what is increasingly being referred to as "The Galanopoulos Thesis." No previous theory concerning the location and nature of Atlantis or the source and manner of its destruction has ever attracted more than a fraction of such support.

seum, his compatriot, George Smith, found several which re-
counted parts of the lost *Epic of Gilgamesh,* a tale embracing the
Babylonian record of the Deluge.[3] On an expedition to Nineveh
underwritten by the *London Daily Telegraph,* Smith had the remarkable good fortune to unearth the missing portions. When added
to the others in an uninterrupted translation their resemblance to
the Hebrew account in Genesis is significant. The following are
pertinent excerpts:

> Man of Shuruppak, son of Ubar-Tutu
> Tear down this house, build a ship!
> Give up possessions, seek thou life.
> Forswear worldly goods and keep the soul alive!
> Aboard the ship take thou the seed of all living things.
> The ship that thou shalt build,
> Her dimensions shall be to measure.
> Equal shall be her width and her length.
> Like the Apsu shalt thou ceil her. . . .

The description of the Deluge is graphic and dramatic:

> Consternation over Adad reaches to the heavens,
> Who turned to blackness all that has been light.
> The wide land was shattered like a pot!
> For one day the south-storm slew,
> Gathering speed as it slew, [submerging the mountains],
> Overtaking the [people] like a battle.
> No one can see his fellow,
> Nor can one be recognized from heaven.

Although, traditionally, Moses has been held to be the author of
Genesis, many scholars now believe it was written centuries after
his time, in fact not earlier than 800 B.C. This may or may not be
so. Regardless of who its creator really was, the fact that he (or
they) drew heavily on the *Epic of Gilgamesh* is made clear by the

---

[3]Carved on clay tablets, many epic tales are to be found in Sumerian literature, none
more remarkable than the Epic of Gilgamesh. This folk hero, ruler over ancient Uruk,
is depicted in such sculpture as the alabaster relief found in the palace of King Sargon
II at Khorsabad.

following passage from the latter, so very like the words of the Old
Testament:

> Mount Nisir held the ship fast, allowing no motion. . . .
> When the seventh day arrived, I sent forth and set free a dove.
> The dove went forth, but came back;
> Since no resting place for it was visible, she turned round.
> Then I sent forth and set free a swallow.
> The swallow went forth, but came back. . . .

Ultimately, the flood was over, whereupon the poem, with an
inexplicable change of tense, recorded: "A raven went forth and,
seeing that the waters had diminished, he eats, circles, caws, and
turns not round."

That Jewish scribes of this period should draw heavily on very
ancient Sumerian sources is not remarkable, nor is it particularly
unusual. For example, it is now almost certain that Solomon's Song
of Songs, that most sensual book of the Bible, was actually taken
from a Canaanite marriage rite peculiar to the Tammuz-Ishtar sect
of the love goddess Astarte, a rite derived in turn from the Dumuzi-
Inanna cult of Sumer. What is significant is that this legend of "a
mighty deluge" appears in the folklore of several ancient cultures,
most especially the Greek, Babylonian, Egyptian, and Hebrew. Thus
the question arises: Were these localized floods taking place at
different periods of time, perhaps even centuries apart? Or do all
the accounts actually speak of the *same* deluge, one universal calamity embracing the entire eastern Mediterranean and the lands
whose coastlines border it? Many scholars today believe the latter
to be the case. Presenting most convincing evidence to identify the
"Great Deluge" of the ancients with the terrible tsunami generated
by the Bronze-Age Santorini eruption, one pair of authorities apply
Aristotelian logic to deal with the more obvious discrepancies of
time and location.

> Lack of rapid communication and the destruction of coastwise
> shipping together with the uniformity of the phenomena could be
> reasons for the flood being considered a local disaster and each of
> the places affected as the starting point of a local deluge rather than
> a common sufferer in a general disaster. This pseudo-local expla-
> nation together with the fact that the Santorin disaster had several

phases could be a reason for the conflicting dates ascribed to the different aspects.[4]

Both Galanopoulos and Bacon on the one hand, and Mavor on the other, make serious attempts to connect the Santorini tsunami with the Israelite exodus from Egypt and the concurrent destruction by drowning of the pursuing army of the pharaoh. They are, of course, not the only authorities to investigate this line of thought. In any case, the concept is so historically intriguing and intellectually challenging as perhaps to bear retelling here.

The actual route out of Egypt taken by Moses and the Israelite host is today entirely uncertain and remains a controversial subject among exegetes. Opinion is divided between two primary possibilities. The majority hold that it began close to the eastern side of the Nile's broad delta, running south down the east coast of the Gulf of Suez, then southeast to Jabal Musa, traditionally identified as Mount Sinai. This is commonly designated "the Southern Route." In its favor is the fact that two springs still to be found along the way are believed by some researchers to be those of Marah and Elim, both Biblical references.

The second possibility, called "the Northern Route," led east along a narrow strip of coastal land which would have put the Mediterranean on the Israelites' left and a broad but shallow lagoon on their right. This route eventually led south to Jabal Hilal, a prominence its advocates say is Sinai. Not far to the east-southeast of this hill it converges with the Southern Route at the oasis of Ayn al Qudayrat, believed by many authorities to be the Kadesh-Barnea of the Bible. It seems fairly certain that the latter was a permanent camp, the Israelite base during much of the forty years of the Wandering. In any case, out of this Wilderness of Zin Moses led his people toward Canaan, reaching the north end of the Dead Sea (then called the Sea of Salt), climbing Mount Nebo to view with dying eyes the Promised Land.

Which of the two was the true way of the Exodus? Despite the fact that for many centuries Greek Orthodox monks have maintained the Monastery of Saint Catherine near the base of Jabal

---

[4]Galanopoulos and Bacon, *Atlantis: The Truth Behind the Legend* (New York: Bobbs-Merrill Company, 1969).

Musa, which for them *is* Sinai, no one knows. If the Biblical manna was, as some believe, an insectival exudation found on Tamarisk trees, the Southern Route is indicated, since this tree grows prolifically in the southern Sinai but only sparsely in the north. Conversely, the Bible says that along the path of flight God provided quail for the Children of Israel, a bird not commonly found in Sinai except on the seacoast. Possibly, there were two separate exoduses, perhaps even years or decades apart, but one thing seems certain. The Israelites of the Exodus did not cross the Red Sea, as is commonly believed. Even if the Southern Route *were* valid, the bulk of available evidence would still support a contention that their main body never went farther south than what is today called the Wilderness of Sinai, a part of the peninsula still many miles north of the Red Sea proper.

At this point of seeming impasse it might prove helpful to call on the discipline of logic. What route would Moses be more likely to have chosen? Let us look at the known facts. The Bible records that the Israelites were slaves of the Egyptians in the Land of Goshen, today believed to lie on the eastern border of the Nile Delta. Then as now, this very flat area, like most regions close to major river estuaries, was well-watered and extremely fertile, excellent for the pasturage of large herds and the raising of crops, close enough to the sea to provide welcome relief from the ever-encroaching desert, the Wilderness of Shur.

Moses knew the region to the south, or at least a part of it. Years before, having killed an Egyptian overseer who was flogging a Hebrew slave, he had fled far to the southeast across the Gulf of Aqaba to the Land of Midian. Having survived to reach that refuge by a miracle he couldn't understand, would he be likely to plunge his people into the nearly waterless, burning hell of southern Sinai? He might do so if that were the only option, but the Northern Route offered another choice. True, warlike tribes of Amorites and Canaanites barred the way, and if the Hebrews kept to the coast they would encounter Gaza, mightiest of Philistine strongholds.[5] But

---

[5]As to which people actually occupied Gaza at the time, scholars remain uncertain. As the Book of Genesis makes clear, even before the calling of Abraham it was a border city of the Canaanites. It is probable that its conquest by the Philistines occurred before the Exodus.

while still to the west of Gaza, they could abandon the coast, as indeed they would have to have done in order to reach Kadesh-Barnea. In the meanwhile, every eastward step they took would bring them closer to Canaan, not farther from it as on the Southern Route.

Why, one must ask, would Moses have made such an extensive detour to the south only to have to come north again through the arid and scorching Wilderness of Paran? To avoid Egyptian camel patrols? Neither the Bible nor Egyptian writings makes any mention of their existence in Sinai, nor would their numbers have been remotely equal to the task of defeating so large a host. But if these arguments for the Northern Route appear to have weight, there is still another more cogent one. Try as one may, one finds nothing along the Southern Route which may remotely be made to fit the story of the parting of the waters of the sea. This miracle is supposed to have occurred near Baal-Zephon at the time the Exodus began. But where was this place? The Biblical scholar, William Smith, places it "on the western shore of the Gulf of Suez, a little below its head. . . ."[6] Other equally eminent researchers believe it may have been located on the Mediterranean coast, somewhere to the east of the Nile Delta between Romani and El Arish.

Now our argument has come full circle. The area between Romani and El Arish is precisely that of the "broad but shallow lagoon" mentioned earlier, that large body of water which would have flanked the Hebrew line of march had they taken the Northern Route. Today, it is difficult to get to by land. Attempting to reach it in 1975, aways blocked by Israeli or Egyptian military positions, one traveler wrote of having to fly from Egypt to Israel via Rome in order to make the only possible approach.[7]

Although the accumulating silt of centuries has considerably extended the Nile Delta since the time of Moses, increasing the amount of arable land, it is the belief of most geological experts on the region that the great lagoon to the east of it (today called in Arabic *Sebcha el Bardawil*) has undergone a contrary process, prob-

---

[6]William Smith, *Smith's Bible Dictionary* (New Jersey: Fleming H. Revell Company, 1967).

[7]See Harvey Arden, "In Search of Moses," *National Geographic*, vol. 149, no. 1 (January 1976).

ably due to earthquake-induced land-slumping and the filling of an adjoining depression by the sea. This is readily substantiated. Writing in the first century B.C., the Greek geographer Strabo recorded the dimensions of the Sirbonis Sea, as it was called then, saying it was 200 *stades* (22 miles) long and 50 *stades* (5½ miles) wide. Today, it is closer to 50 miles in length, with a width near its center of 13 miles. Aerial photos show several small islands, some of which are so low in elevation they would scarcely be visible from the shore. The fact that Strabo and the Sicilian historian Diodorus Siculus both described it as very hazardous to cross, the latter stating that "whole armies" had been lost in it, is a further indication of its extreme shallowness (and, presumably, much smaller size) in ancient times. For what but the existence in those days of vast pockets of quicksand could account for such heavy human casualties? And quicksand means that the doomed soldiers tried to *walk* across. And *that* means it was shallow.

So let us hypothetically suppose that Moses and the Israelites took the Northern Route out of Egypt, moving eastward along the Mediterranean coast in the general direction of Gaza. Very shortly they'd encounter that long strip of sand, today marked by Tel Ras Burun at its widest point, a 120-foot hill beyond which, in those days, the ocean had broken through to merge its waters with the lagoon. The Mediterranean is a nearly tideless sea. Therefore, this channel, called Ekregma (gap), would probably not have offered a vicious current as an obstacle but only the water itself, with the possibility of quicksand along its approaches. Since they were basically farmers and herdsmen, it is logical to presume that most of the Hebrews could not swim. Yet, knowing the well-armed Egyptians to be in rapid pursuit, they must cross this barrier or face annihilation.

This is how it began, apparently, the legend that they crossed the Red Sea. For in later centuries the Sirbonis Sea was known by its Hebrew name of Yam Suph (Jam Suf in Arabic), translated by early geographers of the region as "Red Sea" but now believed to mean "Reed Sea," or literally Sea of Reeds. Seeing this body of water, what thoughts must Moses have had? Today, it is filled in. One could walk the full length of the sandspit, all 50 miles of it, the ocean on the north, the lagoon on the south, no obstacle of any kind. But then? We'll let the Bible tell the story:

And Moses stretched out his hand over the sea; and the Lord caused the sea to go back by a strong east wind all that night, and made the sea dry land, and the waters were divided.

And the children of Israel went into the midst of the sea upon the dry ground: and the waters were a wall unto them on their right hand, and on their left.

So the water of the channel withdrew. We now reach a critical point in our reasoning. If a natural explanation for what occurred is to be advanced, we must assume that the ocean withdrew as well, the Mediterranean itself, shrinking along the shorelines of its eastern regions to rapidly reveal exposed beaches far out. The reader will know what geological condition is implied by this: either a high-magnitude submarine earthquake involving severe crustal slippage; or a volcanic island in eruption, followed by caldera collapse and invasion by the sea; or both. The Book of Exodus continues:

And the Egyptians pursued, and went in after them to the midst of the sea, even all Pharaoh's horses, his chariots, and his horsemen. . . .

And the Lord said unto Moses, "Stretch out thine hand over the sea, that the waters may come again upon the Egyptians, upon their chariots, and upon their horsemen."

And Moses stretched forth his hand over the sea, and the sea returned to his strength when the morning appeared; and the Egyptians fled against it; and the Lord overthrew the Egyptians in the midst of the sea.

And the waters returned, and covered the chariots, and the horsemen, and all the host of Pharaoh that came into the sea after them; there remained not so much as one of them.

Despite the Biblical tendency to imagery and allegory, a tendency that has caused no end of historical confusion, was this an account of a tsunami? More in particular, was it, *could* it have been those tsunami which followed the cataclysmic Bronze-Age eruption of Santorini, waves which destroyed in their ports not only the numerous vessels of the Minoan fleet but, presumably, virtually all shipping then berthed in east Mediterranean harbors? Was it a tsunami? The answer to this first question seems obvious. Despite the allegorical trappings, it has every earmark of being one, a wave

177

or waves, moreover, which followed a protracted series of events typical of eruptive phenomena. Still using the Bible as a reference, let us examine those occurrences which so disastrously fell on the realm of the pharaoh, namely the Ten Plagues of Egypt. Exodus tells us they consisted of the following:

1. All bodies of water change into blood.
2. Frogs appear everywhere on the land.
3. The dust of the land becomes lice infesting man and beast.
4. Swarms of flies appear.
5. A murrain slaughters the livestock.
6. Ashes turn into boils and blains.
7. Hail, thunder, and fire sweep the land.
8. Locusts come on an east wind.
9. Three days of darkness prevail.
10. Death comes to the firstborn of man and beast.

Water into blood. During great eruptions, vast quantities of ash may be carried over long distances by volcanically generated winds, its chemical composition dependent on the minerals in the melt at the time of their explosive diffusion. Frequently, torrential rains may follow such dispersion, the color of the water precipitated varying markedly according to its mineral content. For example, the water of the crater lake found on Java's Mount Kelud is of a deep jade green, the result chiefly of ferrous sulphate ($FeSO_4$), an astringent salt. This is true also of Iceland's Mount Askja crater and Alaska's Mount Mageik, while an infusion of other minerals gives an azure hue to the caldera waters of Alaska's Mount Katmai and Oregon's Crater Lake. How can water turn the color of blood? Ferric oxide (Fe_2O_3), a compound found in nature as hematite, can do that. This mineral has colored a dark ruby red the crater lake of Costa Rica's Mount Irazú and the caldera waters of Anak Krakatoa. The walls of Santorini's caldera are rich in hematite ore. Open to the sea in two places, however, its water is too extensive and diffused to take its color from them.

The rest is as readily explained. In some volcanic eruptions the fantastic heat produces the intense vacuums that may cause tornadoes, such as at Tambora in 1815.[8] If a tornado crosses a lake or river, it will often suck living fish and other freshwater life, including frogs, into its spinning maw, and disgorge them at widely dispersed points. Or the frogs may leave their watery habitat when the accumulating ashfall makes breathing impossible. Both types of occurrences have been recorded in scores of areas throughout the world.

When volcanoes erupt concurrently with high magnitude earthquakes, if the disaster occurs in a heavily populated area its extent may reach horrendous proportions. Together with landslides, crustal subsidence and uplift disrupt the disposition and flow of ground water, creating marshes and swamps, turning formerly fresh water stagnant, making breeding places for harmful insects. Malaria-carrying *Anopheles* mosquitoes may greatly increase their numbers at such times, as may the *Aëdes aegypti* mosquito, carrier of yellow fever. Sewage systems are destroyed, drinking water polluted, corpses left unburied to invite deadly typhus. The settling of great quantities of volcanic dust causes widespread crop failure with its attendant starvation for man and beast. As a concomitant of famine, boils and blains may well appear, while destruction of normal feeding grounds may cause swarms of locusts to migrate unseasonally. And we mustn't forget the ubiquitous rat. Displaced from its habitat by such conditions, the roof rat (*Rattus rattus*) may well play the role it fulfilled during the Middle Ages, that of host animal to such body parasites as the *Xenopsylla* flea, harborer of the bubonic plague bacillus.

Thunder and fire, of course, are commonly associated with volcanic activity, and great hail storms are not at all unusual. Such phenomena may also accompany large earthquakes, including most spectacular pyrotechnical displays, lightning and flood-producing cloudbursts. The Ninth Plague of Egypt was three days of darkness. Clearly, this was caused by volcanic ash. One recalls such accounts as Krakatoa's eruption in 1883 and Katmai's in 1912, just two of many instances where visibility was reduced to zero over many miles from the source of the disturbance.

[8]See Chapter 3.

Lastly, we come to the death of the firstborn. Although Exodus 12, verses 29 and 30, states very clearly that it took place, and although the authorship of the Second Book of the Old Testament is not apocryphal, being generally credited to Moses, it is likely subject to those occasional distortions peculiar to Biblical narrative. At times they are allegorical. At others, the absence of metaphor does not necessarily result in adherence to literal fact. With the Israelites, as with many other peoples, firstborn sons had a special significance. As the climactic and most fearful visitation, the last of the Ten Plagues of Egypt, it is not surprising that the author of Exodus should make this claim in the interest of the ultimate in dramatic value. With regard to Hebrew immunity, one is prone to recall the fact that the Jews were both ethnically and geographically separated from the Egyptians, at least to a large degree. The Land of Goshen was quite effectively isolated.

Giving the above-indicated facts and conjectures their fullest consideration, it now seems feasible to pose the following hypothesis: that the Israelites took the Northern Route during the Exodus; that their flight began after a period of days, weeks, months (or perhaps years), during which time the land of Egypt suffered repeatedly from disasters of volcanic origin; that the source of these catastrophes was Santorini erupting with enormous violence in the not-too-distant Aegean.

There is a school of thought whose advocates consider this supposition to be based on an impossibility. Reminding us that the Bible credits the Hebrews of the Exodus with numbering 600,000 souls, they maintain that so great a host could not possibly have crossed this suddenly waterless gap in the short time between the withdrawal of the sea and the coming of the first tsunami. But other scholars refute this. If you take this figure literally, they claim, then you must accept the rest, that the number was swelled far beyond this by the addition of the descendents of Levi and their multitudinous retainers until the total becomes ridiculous, in no way supportable by the Sinai then or now. The discrepancy is explainable, they say. Today, the Hebrew word *eleph* means "thousand," but there is evidence that in Biblical times it meant "family." Should this be the case, the host of the Children of Israel, far from numbering 600,000 souls or more, comprised only 600 families with their chattel, a total of rather less than 15,000 people. *This* number could have crossed.

To prove that the time interval was sufficient for the Hebrews to have crossed the temporarily drained Ekregma, Galanopoulos and Bacon quote the published research figures of Professor R. Takahasi, President of the Tsunami Committee of the International Union of Geodesy and Geophysics, formerly the director of Tokyo's Earthquake Research Institute, as follows:

> The predominating peaks (in the spectra of tsunami) fall into bands of periods of 84 min and 58 min when the magnitude of the earthquake and accordingly the magnitude of the tsunami is large, while the peaks of the shorter period of 28 min and 21 min are liable to appear when the magnitude of the earthquake is small.

Very clearly, the magnitude of the quakes accompanying the Bronze-Age Santorini eruption was very large indeed. If Takahasi's figures are accurate, therefore, a period of anywhere from an hour to nearly an hour and a half would have passed from the withdrawal of the sea until the arrival of the first wave. Provided a way could be found through the quicksand, this length of time would have been adequate.

Is there Biblical evidence that quicksand was present? The answer is yes. Exodus 14, verses 24 and 25, reads: "And it came to pass, that in the morning watch the Lord looked unto the host of the Egyptians through the pillar of fire and of the cloud, and troubled the host of the Egyptians, and took off their chariot wheels, that they drave them heavily. . . ." This would seem an obvious reference to wheels that were mired.

One of the most arduous tasks facing historians is to fix the dates relating to Israel's captivity in Egypt. The word *Habiru,* a name believed to denote the Hebrews, is rarely referred to in Egyptian chronicles. The pharaoh who ruled in the time of Moses cannot be positively identified. This being the case, is it possible to discover the exact date of the Exodus? Speculation is plagued with pitfalls, particularly in view of the fact that the date of the culminating cataclysm at Thera cannot be pinpointed with any reasonable proof of accuracy. Once again, logic may come to our aid. Did it occur, as is often claimed, during the reign of the warlike Pharaoh Rameses II? He was ruler of Egypt for 68 years, from 1304 to 1237 B.C. However, there is no close correspondence between his dates and that which a consensus of scientific thought now assigns to Santorini's world-shaking outburst.

If not Rameses, who then? After careful consideration, an earlier monarch comes to mind. Pharaoh Thutmose III, stepson-nephew of the enterprising and opportunistic Queen Hat-Shepsut, was Egypt's foremost military figure, considered a genius of war. Coming to the throne on her death in 1481 B.C., he launched the first of his many conquests two years later, defeating a formidable Syrian coalition, then, in successive campaigns lasting two decades, consolidating his rule throughout much of western Asia. Having made his son Amenhotep II Co-Regent, he died in 1447 B.C. And *that* is a date which fits very closely.

It may be argued that the Pharaoh led his army into the area of the fatal crossing, and died with them, or so the Bible seems to imply. Since Thutmose III was buried in the Valley of the Kings and his mummy is now in Cairo, would it be he who was carried out to sea? Ah, but drowned bodies may be washed up on shore, may they not? And many Egyptians were so discovered afterward. Chapter 14 of Exodus, verse 30, reads: "Thus the Lord saved Isreal that day out of the hand of the Egyptians; and Israel saw the Egyptians dead upon the sea shore."

That on the one hand; on the other the fact that the Bible does not specifically state that the Pharaoh died at the crossing, in any case, and presumption is not proof.

Did the plagues and the Exodus extend beyond the length of a single reign? That too is unknown. If they did, it is possible that the pharaohs Amenhotep II and III may somehow have been associated with them.

So there we have the picture, perhaps not as complete as it will one day be, but as accurate as present research would appear to allow. The 300-foot flooding contour has been well-defined. In Lower Egypt it runs far into the desert, spreading across the river delta in a gigantic fan, spanning an arc extending from a point more than 50 miles west of the Nile to beyond the Gulf of Suez and well into the Sinai, 100 miles south of the nearest point on the coast.[9] One thinks of the Galanopoulos-Bacon estimate, tsunami 22 feet high smashing at the ancient port of Jaffa, retaining much of their vast energy after traveling 562 miles. Even a wave of that size would have drowned the Pharaoh's army easily. But shore topography plays a role in the size such waves may attain. What the shelf configuration of the eastern Mediterranean may have been some 35 centuries ago is unknown, nor can it be accurately established. Therefore, the size of the waves which struck Jaffa then are purely speculative. And who can say how big they were when they fell on the passage of the Ekregma?

[9]Due to the funneling contour of a river valley such as that of the Nile, and its gentle gradient, it is logical that such waves would penetrate most deeply in this area. In point of fact, never deviating materially from its mean of 300 feet above sea level, the contour line is still visible *more than 200 miles into Egypt* in the Nile Valley, 120 miles south of the city of Cairo.

12

Can Man Defend Himself?
The Future

The Pacific and Alaskan tsunami warning centers, their methods,
instrumentation, and reach.

It is perhaps fitting that a state which has been struck 41 times
by tsunami since 1819 should today serve as headquarters for a
circum-Pacific detection network designed to give advance warning
against them. That state is Hawaii. The headquarters is called the
Pacific Tsunami Warning Center, or PTWC in abbreviation.[1] Logical, too, is the fact that the network was established by the U.S.
Coast and Geodetic Survey, an organization from whose ranks came
most of its original personnel.[2]

Captain Robert C. Munson was mentioned in Chapter 4. Also of
note is Commander C. K. Green, an oceanographic engineer whose
cleverly designed seismic sea wave detection gauge added considerably to the system's effectiveness. This gauge cannot be activated
by wind waves or the rise and fall of the sea's surface caused by
the daily tides. Distinguishing between the various periods of different types of waves, it automatically rings an alarm only when
the crests are from ten minutes to one hour apart, the intervals
typical of tsunami.

The Green gauge was not in operation in 1946 and the warning

---

[1]Formerly the SSWWS, the Seismic Sea-Wave Warning System. The present head is
Dr. Gordon D. Burton, Geophysicist in Charge.

[2]It is now the National Ocean Survey of the NOAA, the National Oceanic and Atmospheric Administration.

service was not then in existence. The need for such equipment and such an organization was amply demonstrated by the April 1 undersea earthquake in the Aleutian Trench and the resultant heavy damage and loss of life due to strike by tsunami. Already described was the utter destruction of the Scotch Cap Lighthouse on Unimak Island. However, it was primarily the savage blow delivered against Hilo, Hawaii, which prompted the Geodetic Survey to accelerate its plans for a systematized defense.

To anyone who has read this book, driving along the coasts of the island of Oahu would be an adventure in itself. Taking the Kamehameha Highway northward up the coast of Windward Oahu would eventually cause the traveler to swing west to follow the North Shore. Very soon, Kawela Bay would be on the right, still unfaded its memories of Francis Shepard and the tsunami he had the courage to photograph. A little more than 5 miles farther and world famous Waimea Bay and its surfers' Banzai Pipeline would appear, perhaps recalling the size of the winter surf there and the fact that it's nearly as big all the way past Sunset Beach to Kawela. Once beyond Thompson Corner, Farrington Highway would be taken to Kaena Point, then generally south-southeast down the Waianae Coast to Barbers Point and its naval air station. Here the driver would turn eastward. Ahead would lie the entire south coast of Oahu: the vast complex of Pearl Harbor and Hickam Air Force Base, the Honolulu International Airport flanking Nimitz Highway, Honolulu itself, Waikiki Beach, the well-known promontory of Diamond Head.

Between Barbers Point and Pearl Harbor the motorist would pass through the area of Ewa Beach. There would be found the control center of the PTWC, there the personnel and equipment comprising the heart of the tsunami defense system for that vast region of the earth's surface called the Pacific Basin. Note that the site is on the south coast of Oahu, protected from the known routes of tsunami approach by the mass of the island itself.

The system became operative in 1948. Since the vast majority of tsunami are caused by submarine earthquakes, a prime requisite was the establishment of a far-flung network of earthquake-recording units to rapidly locate their epicenters. At this writing (1985), there are 33 seismograph stations in the system. Moving clockwise around the Pacific, a few of them are as follows: Adak, Fairbanks, Palmer, and Sitka in the Aleutians and Alaska; Victoria, British Columbia; Newport, Washington; Berkeley and Pasadena, California; Tucson, Arizona; Huancayo, Peru; Santiago, Chile; Papeete, Tahiti; Apia, Samoa Islands; Guam, Mariana Islands; Manila (Luzon), Philippines; Hong Kong; and Tokyo, Japan.[3]

There is also within the system a second network which complements the first and is equally indispensable. It consists of 53 tide stations, a large percentage of them already operating with the most sophisticated and modern equipment. Again moving clockwise, a sampling would be as follows: Attu, Shemya, and Unalaska in the Aleutians; Cold Bay, Alaska Peninsula; Kodiak Island; Seward and Sitka on the Alaskan Mainland; Tofino, British Columbia; Crescent City, San Pedro, and La Jolla, California; Acajutla, El Salvador; Balboa, Panama; La Punta, Peru; Arica, Antofagasta, Valparaiso, Talcahuano, and Puerto Montt, Chile; Easter Island; Papeete, Tahiti; Christmas Island; Pago Pago and Apia, Samoa Islands; Suva, Fiji Islands; Johnston Island; Kwajalein, Marshall Islands; Wake Island; Guam, Mariana Islands; Legazpi (Luzon), Philippines; Midway Island; Marcus Island (Minami Tori Shima); Shimizu (Shikoku) and Hachinohe (Honshu), Japan.

Put simply, the warning system in the Pacific has two interconnecting phases: first, the detection of an earthquake whose magnitude and location make it potentially tsunamigenic (capable of producing tsunami), and the pinpointing of its epicenter by triangulation; second, confirmation by tide gauges that tsunami actually exist.

When an earthquake in this category has been detected, the Honolulu Observatory (Ewa Beach) issues a tsunami watch. At this time, all members of the tsunami warning system are informed of the quake's occurrence, its precise location, and the fact that tsunami may have been generated. For each member's area, ETAs (estimated times of arrival) of the first wave are computed and given.

The second phase is implemented by tide gauges whose mari-

---

[3]It will be noted that some of these sites, namely Fairbanks, Newport, Pasadena, Tucson, Huancayo and Santiago, are not coastal locations but at some distance inland. This is because an oceanic site is not necessary for a seismograph station even when it is part of the PTWC.

grams or records give the earliest evidence of tsunami generation. Tide stations closest to the epicenter are contacted first. If their gauges have recorded those abnormalities characteristic of a tsunami, the observatory now issues a tsunami *warning*, repeating the ETA's for each location.

PTWC watches and warnings are, of course, advisory only. The alerting and evacuation of populations within the areas of possible strike are the responsibility of local law enforcement, military, and civil defense organizations which over the years have learned to put their faith in the PTWC. It does not issue false alarms. If it issues a tsunami watch, a tsunamigenic earthquake *has* occurred. If this is then upgraded to a tsunami warning, the great waves *have* been generated and are on their way.

The cataclysmic Alaskan earthquake of 1964 had its lessons to teach, the most important of which dealt with the factor of *time*. The killer's speed was shocking, so great that coastal communities near the epicenter of the disturbance could be wiped out long before a warning could be given. It was the realization of this that led to the establishment of the very important regional warning systems for Alaska and the Hawaiian Islands.

Headquartered at Palmer Observatory north of Anchorage, established in 1967, the Alaskan complex extends from Shemya Island at the western extremity of the Aleutian chain all the way to Sitka in the east. Within it are 14 seismograph stations and 8 tide stations. Within it, also, is some of the most sophisticated seismological equipment yet produced by man. Palmer's data-gathering and analysis systems are fully computerized. The robot seismograph and tide stations which feed it information are automated by the most intricate telemetry. The communications network is widespread. Tied into it are all regions of coastal habitation within possible areas of strike; the stations aforementioned, civil defense groups, and military units with assigned emergency duties.

Due to the great vulnerability of many Alaskan settlements, Palmer may issue tsunami warnings not preceded by watches. This does occur with quakes of magnitude 7 or greater, triggering an immediate warning to coastal residents within a 200-mile radius of the epicenter. For those outside that limit a tsunami watch is issued. If incoming tide-gauge reports indicate unusual wave activity, subsequent warnings embrace the full length of the coast.

Set into motion in 1975, the regional system for the state of Hawaii provides for its citizens protection against tsunami of local origin. The nerve center is Honolulu Observatory. Linked with Oahu's quadripartite seismographic network are the stations on Hawaii (the Big Island) and Maui, undersea pressure-sensitive gauges, and four telemetering tide stations.[4]

Honolulu Observatory maintains close and continuous liaison with Hawaii Civil Defense, issuing watch and warning bulletins through this organization. The procedure is similar to the Alaskan System. Since timely warnings alone are of any use, the emphasis in both regional systems is on speed, a factor being constantly and dramatically improved through technological advances.

One of the most impressive of these has been a development of the space age, the geostationary operational environmental satellite, designated GOES, currently maintaining an earth-synchronous orbit 22,300 miles above the equator. The purpose of this marvel of engineering is as simple as its telemetry is complex: to speed communications between far-flung Pacific outposts by providing a permanent orbital station for the instant relaying of vital data. Before its launching, the need for it was aptly phrased by Mark Spaeth, tsunami warning expert attached to the National Weather Service: "We do have serious communications problems getting tide-gauge data out of remote places like Easter Island or the Galapagos. Sometimes we might have to wait a half hour, or even several hours if we're unlucky, for a critical report, depending on the time of day and other traffic. If we were to work through a geostationary satellite, which would require only a single relay point between a remote Pacific Island and NOAA's satellite-acquisition station at Wallops Station, Virginia, this would certainly be a big improvement."[5]

Mark Spaeth got his wish. George Pararas-Carayannis, Director of the ITIC (International Tsunami Information Center) commented as follows in 1982: "Conceptually, based on updated technology and instrumentation, the general operational improvements

---

[4]Honolulu telephone directories contain maps giving dates of previous tsunami strikes of both local and distant origin, and run-up levels in feet for all the main islands.

[5]Quoted from *NOAA*, the quarterly magazine of the National Oceanic and Atmospheric Administration, volume 4, no. 1 (January 1974).

needed to enhance the Tsunami Warning System (TWS) have been developed. The technology of ocean-bottom and shore-based seismic and tsunami sensors transmitting data in real-time to the Pacific Tsunami Warning Center and to regional tsunami centers, making use of GEO-synchronous (GOES) meteorological satellites for communication relay, is now a reality. Future refinements of such improvements may include offshore tsunami bottom sensors using crystal pressure transducers, acoustic or wire links with existing buoy systems, data microprocessors, and data link geostationary satellites between shore installations and the Tsunami Warning Center."[6]

By 1984, the same authority was able to write: "The following scenario evolves: An earthquake activates a seismic instrument. This instrument transmits a signal to the GOES platform, which responds by automatically transmitting an alert code to an alarm device at the warning site designated by local authorities. The alarm device instantly responds by initiating a set of prerecorded instructions based on procedures agreed upon prior to the tsunami. These instructions may include transmitting printed or voice synthesized messages to disaster control headquarters, automatic telephoning of on-duty officials, turning on sirens in population centers, or a wide number of other choices. In addition, GOES also alerts tide gauges near the earthquake to begin sending data via satellite both to local authorities and to PTWC to confirm the presence of a tsunami as a result of the earthquake.

This process, which is entirely automatic, takes no more than 3 minutes to complete. No human decisions are made except possibly for a predetermined one to sound a general alarm.[7]"

Clearly, a great deal has been accomplished since the Honolulu Observatory first became functional in 1948. The effort has become international. The three tsunami warning systems in the Pacific Basin—American, Japanese, and Russian—now share information and coordinate their operations through the Intergovernmental

[6]Quoted from the ITIC Progress Report delivered at Session VIII of the International Coordination Group for the Tsunami Warning System in the Pacific; Suva, Fiji, April 1982.

[7]From the ITIC Progress Report, Session IX of the International Coordination Group for the Tsunami Warning System in the Pacific; Honolulu, Hawaii; 13-17 March, 1984.

Oceanographic Commission, the IOC. This organization is of prime importance. In addition to conducting continuous international liaison among concerned Pacific powers, it maintains (at the Pacific Regional Headquarters of NOAA's National Weather Service) the International Tsunami Information Center. Listed alphabetically, its member nations include Canada, Chile, Ecuador, France, Guatemala, Japan, Korea, New Zealand, Peru, the Philippines, Taiwan, Thailand, the United States, and the Soviet Union.[8]

Much remains to be done. For example, although estimated times of arrival for the first wave of any given wave-train are incredibly accurate, predictions of its magnitude and force cannot as yet be undertaken. The agencies responsible for hurricane watches, dealing with something better known and understood, can issue estimates of wave run-up heights when a storm surge is expected. Not so with tsunami. Not only do topographical variables exist along any coastal region of potential strike, but also at the point of wave origin. In the case of an undersea earthquake, what configuration will the sea bottom take? Will it rise up in a mound without splitting the crust? In this case, theoretically at least, even a very large tsunami will not produce the often-seen abnormal ebb as the sea withdraws from a coast due to be hit. Will the ocean floor slump but fail to fracture? Again, theoretically, no fast suction, no rapid ebb on a given coastline. But suppose the crust splits, as it usually does in a high-magnitude submarine earthquake. Then, depending on the power of the cataclysm, a vast ebb or a comparatively small one may occur along any coast within target range.

What about extra-Pacific tsunami? Since the presently existing warning services in the Pacific are the only organizations of their kind on earth, it is fortunate that the incidence of such waves in other oceans is very much less frequent. Regardless of that, they do occur in those regions; the reader will recall the Caribbean and Mediterranean strikes already noted. Additionally, such instances might be cited as the inundation of the west coast of Calabria, Italy, in 1783 and again in 1908, with thousands drowned on each oc-

---

[8]The IOC is also closely affiliated with the World Data Center A for Tsunami in Boulder, Colorado. This organization bases its operations at the National Geophysical and Solar-Terrestrial Data Center in that city.

casion.[9] In the Atlantic there was Lisbon. In more recent times, on November 18, 1929, the coast of Newfoundland was hit, the causative factor being one of the rarer cases of tsunami propagation, a great submarine landslide. The Arabian Sea produced a tsunami in 1945. On November 27, 1957, its parent body, the Indian Ocean, produced another, a 110-foot wave that swept up from the south to devastate the entire coast of mountainous Baluchistan. There will be others. At present, populations living on the shores of these seas have virtually no protection. Their only salvation lies in providence or blind luck.

If there is one thing of which geophysicists are virtually certain it is that, geologically speaking, Earth will continue in the path of evolution established at the time of its birth, a course immutably set, leading with inexorable logic to the destruction of the planet and its solar system through the death of the sun that created it. Yet astronomers tell us that 5 to 6 billion years may pass before that age of ultimate holocaust arrives, presumably a time when man as we know him will have been long extinct.[10] The process will be a lengthy one. Fifteen billion years may come and go before the sun, having passed through the explosive crisis of metamorphosing into a nova or red giant, becomes an infinitely shrunken, incredibly dense white dwarf, freezing the long-lifeless Earth's surface into a sheet of solid ice.

With events so remote people are not likely to concern themselves, nor even with those much less distant in the immense scale of geologic time. Though in progress as this is written, the ultimate results of the profound changes in the earth's crust due to occur from continental drift and sea-floor spreading are still very far off indeed. One does not fear the onset of the next magnetic reversal,

[9]Both times, the Sicilian port of Messina across the strait of the same name also fell victim.

[10]How is the figure derived? The sun's ratio of hydrogen to helium has been accurately computed, as has the solar conversion rate of the former into the latter through continuous nuclear fusion. When 15 percent of the sun's original hydrogen has been lost the star will enter its nova phase, a lethal prelude to demise lasting about a billion years.

dire though its implications may be; one worries still less about the next ice age.[11] To the nonscientist at least, it is only the present and near future that are to be reckoned with, a future subject to the delimiting factor of a particularly restricted time reference. How long will man survive? Aware of his record as the most savage of all terrestrial predators (the only one that kills for profit and "sport"), most scientists today appear to believe he is destined for relatively early self-extinction, the inevitable result of his moral sense being far outstripped by his technological and military-scientific progress. Be that as it may, his immediate survival is, and quite logically should be, his immediate concern.

Admittedly, the propagation of seismic sea waves is but one of many physical hazards that threaten mankind, and then, except in very rare instances of deep continental penetration, only coastal dwellers.[12] Nonetheless, as the world's population increases at a rate whose dynamic prolificacy even Malthus could not have foreseen, it is inevitable that the numbers of such coastal dwellers will greatly increase. People in this situation should bear the following in mind. The existence of enormous submarine trenches is not a theory but a proven fact. The hypothesis of global tectonics and its corollary of plate subduction, like that of continental drift, is now buttressed by a wealth of supportive evidence amounting to virtual substantiation. If subduction continues, and it will, earthquake-caused tsunami will be periodically generated. If vulcanism continues, and it will, they will also occur, though the incidence of such waves from this cause and from submarine landslides will remain substantially less.

Very obviously, a warning system for the Atlantic and Indian Oceans should be organized, and operations continually improved at existing Pacific facilities. Perhaps of nearly equal importance,

---

[11]So far as concerns reversals of the earth's magnetic field, they occur every half million years on the average and are signaled by a steady drop in magnetic intensity. According to Lamont Geological Observatory's Dr. Bruce Heezen, speaking of the present reduction rate before the International Oceanographic Congress in Moscow, "If this drop continues for 2,000 years, it will be zero." Quite simply, he was saying that the next reversal is nearly due. Collapsing the protective ionosphere, it would expose the Earth to deadly cosmic-ray radiation from outer space.

[12]See references to waves generated by Santorini in Chapter 11.

however, is the lessening of worldwide ignorance of this subject. Remember the Philippines, August 23, 1976?[13] At a time when the international scientific community knew much about tsunami behavior, very little information had been disseminated to this area or, for that matter, to most others of potential risk. The result was a deadly surprise for a large and helpless population and a death toll approaching 8,000. Such calamities, though often avoidable, will happen again. It is the author's hope that this book may be instrumental in reducing their number.

[13]See the Introduction.

Bibliography

Arden, Harvey. "In Search of Moses." *National Geographic,* vol. 149, no. 1, January 1976.

Behrman, Daniel. *The New World of the Oceans.* Boston and Toronto: Little, Brown & Company, 1969.

Bernstein, J. "Giant Waves." *The World of Geology.* Edited by L. D. and F. Leet. New York: 1961.

Blanchard, Duncan C. *From Raindrops to Volcanoes.* New York: Doubleday & Company, Inc., 1966.

Brown, Billye W. and Brown, Walter R. *Historical Catastrophes: Earthquakes.* Massachusetts: Addison-Wesley Publishing Company, Inc., 1974.

Calder, Nigel. *The Restless Earth: A Report on the New Geology.* New York: Viking Press, 1972.

Carrington, Richard. *A Biography of the Sea.* New York: Basic Books, Inc., 1960.

Carson, Rachel. *The Sea Around Us.* New York: Oxford University Press, 1961.

Chapman, Sidney. *IGY: Year of Discovery.* Ann Arbor, Michigan: The University of Michigan Press, 1959.

Cottrell, Leonard. *The Bull of Minos.* New York: Rinehart & Company, Inc., 1958.

Cowen, Robert. *Frontiers of the Sea.* New York: Doubleday & Company, Inc., 1960.

Deacon, G. E. R., ed. *Oceans—An Atlas-History of Man's Exploration of the Deep.* London: Paul Hamlyn, Ltd., 1962.

Desborough, V. R. D'A. *The Greek Dark Ages.* New York: St. Martin's Press, 1972.

Donn, L., and Ewing, M. "A Theory of Ice Ages III." *Science,* vol. 152, 1966.

Evans, Sir Arthur. *The Palace of Minos.* Cheshire, CT: Biblo and Tannen Booksellers, 1921.

Fried, John J. *Life Along the San Andreas Fault.* New York: Saturday Review Press, 1973.

Bibliography

Galanopoulos, Angelos. On the Origin of the Deluge of Deucalion and the Myth of Atlantis. *Greek Archaeological Society,* vol. 3, 1960.

Galanopoulos, Angelos, and Bacon, Edward. *Atlantis: The Truth Behind the Legend.* New York: Bobbs-Merrill Company, 1969.

Gaskell, T. F. *World Beneath the Oceans.* New York: Natural History Press, 1964.

Graves, William, ed. *Hawaii.* Washington, D.C.: National Geographic Society, 1970.

Grote, George. *A History of Greece.* New York: AMS Press, Inc., 1975.

Halacy, D. S. *Earthquakes—A Natural History.* New York: Bobbs-Merrill Company, 1974.

Hallam, A. *A Revolution in the Earth Sciences.* Oxford: Clarendon Press, 1973.

Hedervari, P. Volcanophysical Investigations Concerning the Energetics of the Minoan Eruption of Volcano Santorin. *Bulletin Volcanologique,* vol. 25, 1967.

Heezen, B., and Ewing, M. "The Mid-Oceanic Ridge and Its Extension Through the Arctic Basin." *Geology of the Arctic.* Toronto, 1961.

Heintze, Carl. *The Circle of Fire.* New York: Hawthorn Books, Inc., 1970.

Hewitt, R. *From Earthquake, Fire and Flood.* New York: Charles Scribner's Sons, 1957.

Hodgson, John. *Earthquakes and Earth Structure.* Englewood Cliffs, New Jersey: Prentice-Hall, Inc., 1964.

Idyll, C. P., ed. *Exploring the Ocean World.* New York: Thomas Y. Crowell Company, Inc., 1969.

Imamura, Akitune. *Theoretical and Applied Seismology.* Tokyo: Maruzen Company, 1937.

Ingmanson, Dale E. and Wallace, William J. *Oceanology: An Introduction.* Belmont, California: Wadsworth Publishing Company, 1973.

Lane, Frank W. *The Elements Rage.* Pennsylvania and New York: Chilton Books, Inc., 1965.

Leet, L. Don, and Judson, Sheldon. *Physical Geology.* Englewood Cliffs, New Jersey: Prentice-Hall, Inc., 1965.

Macdonald, Gordon, and Abbott, Agatin. *Volcanoes in the Sea: The Geology of Hawaii,* Honolulu: University of Hawaii Press, 1970.

Marinatos, Spyridon. "About the Rumour of Atlantis." *Cretika Chronika,* vol. 4., Heraklion, 1950.

————. "The Volcanic Destruction of Minoan Crete." *Antiquity,* vol. 13, 1939; *Crete and Mycenaean Greece.* Athens, 1959.

Matthews, S. W. "The Night the Mountains Moved." *National Geographic,* vol. 117, no. 3, March 1980.

Mavor, James W., Jr. "A Mighty Bronze Age Volcanic Explosion." *Oceanus,* vol. 12, no. 3, April 1966.

Bibliography

————. *Voyage to Atlantis*. New York: G. P. Putnam's Sons, 1969.

McKee, Bates. "The Restless Edge." *Our Continent: A Natural History of North America*. Washington, D.C.: National Geographic Society, 1976.

Miller, Donald. "The Alaska Earthquake of July 10, 1958, Giant Wave in Lituya Bay." *Bulletin of the Seismological Society of America*, vol. 50, 1960.

Ninkovich, D., and Heezen, B. "Santorini Tephra." *Colston Papers*, vol. 17. London: Butterworth Scientific Publications, 1965.

Palmer, L. T. *Mycenaeans and Minoans*. New York: Alfred A. Knopf, Inc., 1962.

Pararas-Carayannis, George. International Tsunami Information Center (ITIC) *A Progress Report for 1980–1982*. Session VIII of the International Coordination Group for the Tsunami Warning System in the Pacific. Suva, Fiji, April 1982.

————. International Tsunami Information Center (ITIC) *A Progress Report for 1982–1984*. Session IX of the International Coordination Group for the Tsunami Warning System in the Pacific. Honolulu, Hawaii, March 1984.

Pirie, R. Gordon, ed. *Oceanography: Contemporary Readings in Ocean Sciences*. New York: Oxford University Press, 1977.

Press, F., and Harkrider D. "Air-Sea Waves from the Explosion of Krakatoa." *Science*, vol. 154, 1966.

Putnam, William. *Geology*. London and New York: Oxford University Press, 1964.

Richter, C. F. *Elementary Seismology*. San Francisco: W. H. Freeman and Company, 1958.

Rossi, Frank P. "Seaquakes: Shakers of Ships." *Mariners Weather Log*, vol. 2, no. 5, September 1967.

Saint-Amand, Pierre. "Los Terremotos de Mayo, Chile, 1960." China Lake, California: Michelson Laboratories, U. S. Naval Ordnance Test Station, *Technical Publication 2701*, August 1961.

Shelton, John S. *Geology Illustrated*. New York: W. H. Freeman and Company, 1966.

Shepard, Francis P. *The Earth Beneath the Sea*. New York: Atheneum Publishers, 1963.

Smith, F. G. Walton. *The Seas in Motion*. New York: Thomas Y. Crowell Company, 1973.

Smith, William. *Smith's Bible Dictionary*. New Jersey: Fleming H. Revell Company, Inc., 1967.

Sullivan, Walter. *Assault on the Unknown*. New York: McGraw-Hill Book Company, 1961.

Sverdrup, Harold U., Johnson, Martin, W., and Fleming, Richard H. *The Oceans*. New Jersey: Prentice-Hall, Inc., 1942.

Takahasi, R., and Aida, I. "Studies of the Spectrum of Tsunamis." *Bulletin of the Earthquake Research Institute*, vol. 39, Tokyo, 1961.

Bibliography

Tazieff, Haroun. *The Orion Book of Volcanoes*. New York: The Orion Press, 1961.

————. *When the Earth Trembles*. New York: Harcourt, Brace & World, Inc., 1964.

————. *The Making of the Earth*. Westmead, Farnsborough, Hants, England: Saxon House, D. C. Heath, Ltd., 1974.

Wilcoxson, Kent. *Chains of Fire: The Story of Volcanoes*. Philadelphia and New York: Chilton Books, Inc., 1966.

Wyckoff, Jerome. *The Story of Geology*. New York: Golden Press, 1960.

Yoshio, Kato, et al. "The Chile Tsunami of 1960 Observed Along the Sanriku Coast of Japan." Science Report, Tôhoku University 5: *Geophysics,* vol. 13 (1961).

Zetler, Bernard D. "The Seismic Sea-Wave Warning System: An Aid to Mariners." *Mariners Weather Log*, vol. 9, no. 5, September 1965.

Index

Abyssal plains, 16
Aconcagua ("Butcher of the Andes"), 15
Alaska:
  Denali Fault System in, 77–79
  earthquake (1964) in, 80–82
  Lituya Bay, 144–146
    swash (1958) in, 146–148
  tidal bore in, 136
  tsunami:
    1946 (Unimak Island), 47–54, 63
    1964, 82–96
    warning system, 188
Aleutian Islands:
  fault system of, 78, 79
  Unimak Island tsunami (1946) at, 47–54
Aleutian Trench, 19
American Plate, 67, 70, 112–113
Amundsen, Roald, 145n
Anchorage (Alaska), 79
  earthquake (1964) at, 81–82
Anjer Lor (Java), 32, 33
Arabian Sea, 192
Arica (Chile), tsunami (1868) at, 39–43, 55–58, 62
Armstrong, Karl, 85
Assam Earthquake, 134n
Asthenosphere, 65–66
Atlantic Ocean:
  Azores-Gibraltar Ridge, 119–120
  and continental drift process, 24

Iberian Basin, 119
Mid-Atlantic Ridge, 14, 21, 24
Puerto Rican Trench, 113, 118–119
tsunami in, 192
  Lisbon, Portugal (1755), 124–125
Atlantis, site of, 170–171
Aubert, Jim, 90, 91
Azores-Gibraltar Ridge, 119–120

Bacon, Edward, 161, 162–163, 166, 169, 170, 173, 183
Barnesky, Al, 118n
Basins, natural oscillatory period for, 129–130
Bay of Fundy, Nova Scotia, tidal bore in, 135–136
Behrman, Daniel, 160, 161, 165
Berouw (Dutch warship); in Krakatoa tsunami, 33, 39
Bezimianny volcano, 131n
Bible, tsunami in:
  Exodus from Egypt, 176–177, 180–183
  Flood, 170–172
  Ten Plagues of Egypt, 178–180
Billings, L.G., 40, 57, 62
Blegen, Carl W., 152
Bores, tidal, 134–137
Bristol Channel, tidal bore in, 136
Burton, Dr. Gordon D., 185n

Cádiz (Spain), tsunami (1755) at, 127
California, San Andreas Fault in, 18, 66–67, 77, 78
Cape Lopatka, Kamchatka Peninsula, 37, 62
Caribbean:
  earthquakes in, 113
  fracture zone of, 113
  Sunken ships in, 111–112
  tsunami at, 113–114
    Port Royal (1692), 114–118
    Saba (1755), 128
Caribbean Plate, 112–113
Chadwick, John, 152n
Chains of Fire: The Story of Volcanoes (Wilcoxson), 44
Challenger, H.M.S., 13–14
Challenger Deep, 17, 19, 98
Chile:
  earthquake (1960) at, 36, 70–72
  tsunami at:
    1868 (Arica), 39–43, 57–58, 62
    1960, 36, 72–75
  volcanic and seismic instability of, 68–70
China, Fuch'un bore in, 135
Clive, Robert, 136
Cocos Plate, 67, 112
Colombian Basin, 113
Colvin, Jack, 48, 53
Concepción (Chile), 39, 68, 70
Continental drift theory, 21–23, 24, 65, 77
Continental shelf, 14–15
Continental slope, 14, 15–16
Continents, formation of, 22–24, 65–68
Contraction theory, of continent building, 23
Convection current theory, 20–21, 23–24, 66
Cook, James, 57, 145
Cook Inlet, Alaska, 79, 82, 83, 95, 136

Cook Inlet Fault, 78, 79
Core of earth, 6–7
Crete, destruction of Minoan Empire on, 150, 152–168
Crust of earth, 5–6
  changes in, 21–24, 65–68, 77
Cuthbert, Bill, 85–86, 87

Deep-ocean basin, 14, 16
  (*See also* Pacific Ocean Basin)
Deluge (*see* Flood)
Denali Fault System, 77–79, 144
Dilatancy phenomenon, 80
Discontinuities, 10
Dykstra, Dewey, 48, 53

Earth:
  core, 6–7
  crust, 5–6
  crustal change theory:
    continental drift, 21–23, 24, 65, 77
    contraction, 23
    convection current, 20–21, 23–24, 66
    plate movement, 65–68
  mantle, 6
  at sun's death, 192–193
  (*See also* Ocean floor)
Earth Beneath the Sea, The (Shepard), 54–56
Earthquakes:
  Alaska, 79
    1964, 80–82
  casualties in, 71, 98n
  Chile (1960), 70–72
  crustal displacement in, 106–108
  displays of luminescence in, 63
  epicenter, 10–11
  in island arcs, 99
  Japan, 98–100, 101
    1891 (Mino-Owari), 102–103
    1923, 105–107

1930, 108
1933, 108–109
Lisbon, Portugal (1755), 120–124,
    125–128, 132, 141
in Pacific Basin, 18
recording network for, 186–
    187
seiches induced by, 131–134
seismic waves generated by, 140–
    144
seismological investigation of, 7–
    11, 186–187
shock waves of, 8–10, 131–132,
    140
swash generated by, 147–148
in tsunami formation, 2, 35,
    96
East Pacific Rise; 21
Echo sounder, 14
Egypt (ancient):
Hebrew Exodus from, 176–177,
    180–183
Ten Plagues of, 178–180
English Channel, 135–136
Epic of Gilgamesh, 171
Evans, Arthur, 152
Evans, Barry, 104n
Ewing, Maurice, 23–24
Exodus (Biblical):
date of, 181–182
route of, 173–176
Santorini tsunami and, 176–177,
    180–183

Fairweather Fault, 78–79, 81n,
    144–145, 148
Fire storms, 106
Flood (Deluge):
accounts of, 170–172
Santorini tsunami as, 172
Fouqué, Ferdinand, 156–157
Fracture zones, 66, 77
of Pacific Basin, 18–19, 67, 69
Fuch'un bore, 135

Galanopoulos, Angelos, 159n, 160,
    161, 162–163, 166, 170, 173,
    183
Genesis, account of Flood in, 171–
    172
Geostationary operational environ-
    mental satellite (GOES), 189,
    190
Graves, William, 59
Gravimetric prospecting, 99n
Great Java Trench, 19
Great Rift Valley of East Africa,
    24
Greece (ancient):
Minoan Empire, 152–168
Troy, 151–152
Green, C.K., 185
Green gauge, 185
Grote, George, 151n
Gutenberg, Beno, 148
Gutenberg discontinuity, 10
Guyots, 21

Halacy, D. S., 165–166
Hardin, Bob, 90, 91
Harkrider, David, 166
Hawaiian Islands, tsunami in, 57,
    59–60
1946, 54–57
1960 (Hilo), 32, 38, 59, 76
warning center for, 185, 186–
    189
Hebgen Lake, Montana, 133–134
Hebrew Exodus from Egypt, 176–
    177, 180–183
Heezen, Bruce, 23, 24, 160, 162,
    193n
Hewitt, R., 107–108
Hilo (Hawaii):
tsunami at, 32, 38, 57, 59, 76
vulnerable location of, 60
Homer, 151
Homer (Alaska), 95
Honolulu Observatory, 187, 189

Hooghly River, tidal bore of, 136
Hydraulic gradient, 137

Iberian Basin, 119
Indian Ocean, 192
Intergovernmental Oceanographic
    Commission (IOC), 190–191
International Tsunami Information
    Center (ITIC), 189–190
Isostatic equilibrium, 20

Jamaica, Port Royal earthquake-
    tsunami (1692) in, 114–118
Japan:
    earthquakes in, 98–100
    1891 (Mino-Owari), 102–103
    1923, 105, 107
    1930, 108
    1933, 108–109
    fault system of, 100–101
    trench system of, 97–98, 99
    tsunami at, 100
    1854, 101–102
    1896, 103–105
    1933, 63–64
    1960, 76
Java, Krakatoa eruption-tsunami
    (1883) at, 27–34, 161, 165–
    166, 167, 168
Johns Hopkins Glacier, Alaska, 144,
    146

Kaguyak (Alaska), 84, 88–89
Kamchatka Peninsula, 37, 62, 131n
Kenai Peninsula, Alaska, 78, 79, 82,
    83, 92, 95
Knossos (Minoan Empire), 152, 165
Kodiak Island, Alaska, 79, 80, 84–
    88
Krakatoa:
    Anak Krakatoa, 45
    volcanic eruption (1883) at, 27–
    31, 161

tsunami of, 29–30, 31–34,
    165–166, 167, 168

La Pérouse, Jean François de, 145–
    146
Lake Clark Fault, 78
Layard, Austen Henry, 170–171
Link, Edwin, 117
Lisbon (Portugal), earthquake-tsun-
    ami (1755) at, 120–128, 132,
    141
Lituya Bay, Alaska, 79n
    exploration of, 144–146
    swash (1958) in, 146–148
Longitudinal waves, 8–9, 140
Luminescence, tsunami, 62–64

Madeira Islands, 127
Maelstrom (*see* Moskenstraumen)
Mamet-Gorceix excavations, 156,
    159
Mantle of earth, 6
Mariana Trench, 98
Marinatos, Spyridon, 157–158, 165,
    169
Marx, Robert, 117
Mascaret of river Seine, 136–137
Matsubara, Shigeto, 59
Mavor, James, 159, 161, 163–164,
    166, 167, 169, 170, 173
Mediterranean:
    seismic instability in, 149–150
    tsunami in, 191
Mendocino Escarpment (Fracture
    Zone), 18–19, 67
Merak (Java), 33, 37
Mexico, earthquake (1932) in, 142–
    143
Michel, John, 132
Microseisms, 8
Mid-Atlantic Ridge, 14, 21, 24, 67,
    119, 120
Middle America Trench, 19
Mid-Oceanic Ridge, 21, 24, 77

Millsap, Darrell, 53–54, 63
Milne, John, 7
Minoan Empire, destruction of, 150, 152–168
Mino-Owari earthquake, Japan (1891), 102–103
Mohorovičić, Andrija, 10
Mohorovičić, discontinuity (Moho), 10
Mont-Saint-Michel, 136n
Moore, Osborne, 135
Morgan, Henry, 114, 117
Morocco, tsunami (1755) at, 127
Moses, 171, 173, 174, 175, 177
Moskenstraumen, 138–139
Mount Fairweather, Alaska, 145
Mount Ida, Crete, 167
Mount La Pérouse, Alaska, 145n
Mountains, creation of, 68
Munson, Robert C., 59–60, 185

National Oceanic and Atmospheric Administration (NOAA), 189, 191
Nazca Plate, 67–68, 70, 77
Nelson, Horatio, 118
Neo Fault, 102–103
Ness, Paul, 48, 53
New Madrid (Missouri), 132–133
Newfoundland, 192
Niagara Falls, 138
Ninkovich, Dragoslav, 160–161, 162
Noctiluca miliaris, 64
Northwest Passage, 145n

Oahu, Hawaii, tsunami (1946) at, 54–56
Ocean floor:
continental shelf, 14–15
continental slope, 15–16
deep-ocean basin, 16–17
fracture zones of, 18–19, 67–68, 77
mountains of, 21

trench systems of, 19–21, 67, 77, 97–99
vulcanicity in, 17–18
earthquake displacement on, 107–108
soundings of, 13–14
Oceans, The: Their Physics, Chemistry, and General Biology (Sverdrup, Johnson and Fleming), 44
Omori, Fusakichi, 106
Oscillations:
natural tidal, 129–130
nontidal (*see* Seiches)

P (primary) waves, 8–9, 140
Pacific Ocean Basin:
fracture zones of, 18–19, 67, 69
mountains in, 21
trench systems of, 19–21, 67, 97–99
tsunami defense system for, 186–191
volcanic activity in, 17–18
Pacific Plate, 67, 77, 113
Pacific Tsunami Warning Center (PTWC), 75, 95, 185–188, 190
Palmer Observatory, Alaska, 188
Pararas-Carayannis, George, 189–190
Pentland Firth (Swilkie), 138
Peru-Chile Trench, 19, 67, 69, 70, 77
Pettit, Anthony, 48, 53
Philippines, tsunami (1976) in, 1
Phipps, Constantine, 13
Piccard, Jacques, 98n
Pickering, Leonard, 48, 53
Plagues of Egypt (Biblical), tsunami at, 178–180
Plate tectonics theory, 65–68, 77
Plunging wave, 58–59
Poe, Edgar Allan, 138
Pombal, Sebastião de Carvalho, 121, 125–127

Port Royal (Jamaica), earthquake-tsunami (1692) in, 114–118
Portugal, Lisbon earthquake-tsunami (1755) in, 120–128, 132, 141
Press, Frank, 166
Prince William Sound, Alaska, 78, 79, 83, 95–96
Puerto Rico Trench, 113, 118–119
Puyéhue volcano, Chile, 69, 72

Red Sea:
    in continental drift process, 24
    Exodus of Hebrews, tsunami in, 176–177, 180–183
Reloncavi fault, 69
"Ring of Fire," 17–18
Ross, James Clark, 13

Saba, tsunami (1755) at, 128
Sabine, Edward, 22
Sabine Islands, 22
Sagami Nada (Gulf of Sagami), Japan, 101, 105–107
Saint-Amand, Pierre, 69
San Andreas Fault, 18, 66–67, 77, 78
Sanriku Coast, Japan, 76, 98, 100
    earthquake (1933) on, 108–109
    tsunami (1896) on, 103–105
Santorini eruption-tsunami (Bronze Age):
    in destruction of Minoan Empire, 154–168
    in Exodus from Egypt, 176–177, 180–183
    as Flood, 170–172
    and lost Atlantis, 169–170
    in Ten Plagues of Egypt, 178–180
Santorini islands:
    volcanic activity in, 154–155
    (See also Santorini eruption-tsunami)
Sapper, Karl, 17

Schliemann, Heinrich, 151, 152
Scotch Cap Lighthouse, Unimak Island, destruction of, 47–54
Scylla and Charybdis, legend of, 137
Sea bottom (see Ocean floor)
Seamounts, 21
Seaquake, 140
Seiches:
    defined, 128, 130–131
    earthquake-induced, 131–134
Seine River, tidal bore of, 136–137
Seismic sea wave (see Tsunami)
Seismic Sea-Wave Warning System [SSWWS] (see Pacific Tsunami Warning Center [PTWC])
Seismic waves, 140–144
Seismogram, 9
Seismographs:
    determination of epicenter, 10–11
    pendulum principle in, 7–8
    recording shock waves, 8–9
    sensitivity of, 8
    stations, 186–187
Seismology, history of, 7, 126
Seldovia (Alaska), 95
Seldovia Fault, 78
Seward (Alaska), 92–94
Shear waves, 9
Shepard, Francis P., 54–56
Shimoda (Japan), 105n
Shock waves:
    discontinuity, 10
    recording, 9, 186–187
    seismic, 140–144
    types of, 8–9
Sial (crust), 5
Sima (crust), 6
Smith, George, 171
South America:
    Nazca Plate of, 67–68, 70, 77
    Pacific Coast, tsunami strikes on, 39, 68
    (See also Chile)
Spaeth, Mark, 189
Spilling wave, 58
Sumatra, tsunami (1883) at, 33–34

Sun, death of, 192–193
Swash, in Lituya Bay, Alaska, 146–148

Takahasi, R., 181
Tazieff, Haroun, 120n, 154
Teluk Betung (Sumatra), 32, 33
Thera, island of, volcanic eruption on (*see* Santorini eruption-tsunami)
Thutmose III, 160, 182
Tidal bores, 134–137
Tidal wave (*see* Tsunami)
Tide:
    gauges, 185, 187
    semidiurnal, 136n
    stations, 187, 188
Titanic, 144n
Tocantins bore, 135n
Tokyo (Japan), earthquake (1923) at, 106
Transverse waves, 9
Trench systems, 66, 77
    of Pacific Basin, 19–21, 67, 97–99
Troy, lost city of, 151–152
Tsunami (seismic sea waves):
    advance warning systems for, 185–192
    Alaska:
        1946 (Unimak Island), 47–54
        1964, 82–96
    in ancient world (*see* Santorini eruption-tsunami)
    Caribbean (1692), 113–118
    causes of, 2–3, 35
        earthquakes, 2, 35, 96
        volcanic eruptions, 29–30, 38–39
    Chile:
        1868 (Arica), 39–43, 57–58, 62
        1960, 72–75
    Hawaiian Islands, 57, 59–60
        1946, 54–57
        1960, 32, 38, 59, 76

Japan, 100
    1854, 101–102
    1896, 103–105
    1933, 109
    1960, 76
    Krakatoa (1883), 29–30, 31, 165–166, 167, 168
    Lisbon, Portugal (1755), 124–125
    lowering of tide at, 37–38
    ocean movement of, 35, 44–45
    Portugal (1755), 124–125
    power of, 60–61, 62
    versus sea waves, 2
    successive waves of, 35, 38, 56
    term, 1–2
    terminal modification of, 36–37
    Unimak Island, Alaska (1946), 47–54
    variance of form in, 57–59
    velocity of, 35–36
    wave luminescence in, 62–64
Tsunami Warning System (TWS), 189–190
Turbidity currents, 16
Tuscarora Deep, 98

Ulrich, Howard, 146–147
"Ultimate wave, the," 147
Unimak Island, Alaska, tsunami (1946) at, 47–54, 63

Valdez (Alaska), 89–92
Valdivia (Chile), 39, 68, 71
Van Gestle (vulcanologist), witness to Krakatoa eruption, 27–29, 30
Venezuela Basin, 113
Ventris, Michael, 152n
Victoria Falls, 138
Vityaz Deep, 98n
Volcanic eruptions:
    in ancient world (*see* Santorini eruption-tsunami)
    Krakatoa (1883), 27–31, 161

Volcanic eruptions (*cont.*):
 in Pacific Basin, 17–18
 in tsunami formation, 2, 29–30,
  38–39
Volcanic islands, creation of, 39

Wateree, U.S.S., in Chilean tsunami, 39–43, 58
Waves:
 natural tidal oscillation, 129–130
 nontidal (*see* Seiches)
 power of storm waves, 61–
  62
 sea waves, 2, 37
 seismic sea waves (*see* Tsunami)
 spiller and plungers, 58–59

swash, 146–148
tidal bores, 134–137
whirlpools, 137–139
Wegener, Alfred, continental drift
 theory of, 21–23, 65
Whalen, Bernard, 90
When the Earth Trembles (Tazieff),
 120n
Whirlpools, 137–139
Wilcoxson, Kent, 44

Yucatan Basin, 113

Zambesi River, whirlpool of, 138n
Zones of shear, 3